Springer Theses

Recognizing Outstanding Ph.D. Research

Aims and Scope

The series "Springer Theses" brings together a selection of the very best Ph.D. theses from around the world and across the physical sciences. Nominated and endorsed by two recognized specialists, each published volume has been selected for its scientific excellence and the high impact of its contents for the pertinent field of research. For greater accessibility to non-specialists, the published versions include an extended introduction, as well as a foreword by the student's supervisor explaining the special relevance of the work for the field. As a whole, the series will provide a valuable resource both for newcomers to the research fields described, and for other scientists seeking detailed background information on special questions. Finally, it provides an accredited documentation of the valuable contributions made by today's younger generation of scientists.

Theses may be nominated for publication in this series by heads of department at internationally leading universities or institutes and should fulfill all of the following criteria

- They must be written in good English.
- The topic should fall within the confines of Chemistry, Physics, Earth Sciences, Engineering and related interdisciplinary fields such as Materials, Nanoscience, Chemical Engineering, Complex Systems and Biophysics.
- The work reported in the thesis must represent a significant scientific advance.
- If the thesis includes previously published material, permission to reproduce this must be gained from the respective copyright holder (a maximum 30% of the thesis should be a verbatim reproduction from the author's previous publications).
- They must have been examined and passed during the 12 months prior to nomination.
- Each thesis should include a foreword by the supervisor outlining the significance of its content.
- The theses should have a clearly defined structure including an introduction accessible to new PhD students and scientists not expert in the relevant field.

Indexed by zbMATH.

Kon H. Leung

The Strontium Molecular Lattice Clock

Vibrational Spectroscopy with Hertz-Level Accuracy

Springer

Kon H. Leung
California Institute of Technology
Pasadena, CA, USA

ISSN 2190-5053 ISSN 2190-5061 (electronic)
Springer Theses
ISBN 978-3-031-47646-4 ISBN 978-3-031-47647-1 (eBook)
https://doi.org/10.1007/978-3-031-47647-1

This Springer imprint is published by the registered company Springer Nature Switzerland AG
The registered company address is: Gewerbestrasse 11, 6330 Cham, Switzerland

This book is dedicated to the memory of Tota the Cute Turtle,

and the memories of all who have been a part of it.

Supervisor's Foreword

Laser cooling techniques allow scientists to reach temperatures that are orders of magnitude colder than those of any naturally occurring matter. Developed for atoms, these techniques have recently been expanded to small molecules, both via direct application and by building the molecules from laser-cooled atoms. These ultracold systems have been used with enormous success in experiments that have elucidated a variety of quantum effects as well as in highly precise measurements that are done on a table-top. This class of experiments includes atomic clocks and allows physicists to carry out high-precision measurements at ultralow energies with an impact on fundamental science.

Atomic clocks have undergone an unprecedented period of growth thanks to a combination of laser cooling and trapping, extreme laser stabilization, and femtosecond frequency combs. The ability to create and control molecules near absolute zero relies on extending the atomic laser cooling and manipulation techniques to more complex systems. Kon Leung's thesis work is a tour de force demonstration of how all ingredients must come together in order to create clocks where the ticking rate is based on molecular vibrations rather than electronic excitations. It marries the challenges of creating a new type of diatomic molecules in the microkelvin temperature regime and manipulating the molecular quantum states, with world-class metrological tools. This achievement has required Kon to exhibit the creativity of controlling a fragile quantum system alongside showing deliberate patience and care of studying minute systematic effects that could affect the molecular clock.

When Kon joined the laboratory, he found that we had the initial capability to carry out high-precision spectroscopic studies on ultracold diatomic strontium molecules that are held in optical lattices, or interference patterns of laser light. These molecules were created from trapped clouds of laser-cooled atoms and, as such, were barely bound. It was possible to place these weakly bound molecules into coherent quantum-state superpositions, but to make an impact in metrology and fundamental physics, Kon took pioneering steps to transfer the molecules to their lowest energy states, and to engineer their trapping environment to achieve very long-lasting quantum state superpositions. This enduring coherence ultimately allowed Kon to demonstrate quality factors for the molecular clock of several trillion.

The high-quality molecular clock resonances directly translate into an ability to fully understand and quantify this unique system for the first time. A measurement of time or frequency is meaningless if one is not able to quote its uncertainty. The clock's ticking rate can be affected by various factors that include, for example, optical fields applied in the laboratory to hold and probe the atoms or molecules, as well as ambient environmental factors such as thermal radiation from the trapping chamber and stray electromagnetic fields.

In case of the vibrational molecular lattice clock, Kon took up the task to enumerate all the conceivable effects, and to perform painstaking measurements of their magnitudes. Any system studied for the first time at a new level of precision is bound to uncover something unexpected. In particular, Kon discovered that the uncertainty budget of the molecular clock is dominated by the presence of the trapping light. Moreover, for the first time he clearly observed and quantified hyperpolarizability of pure vibrational states in nonpolar molecules, which is a nonlinear response of the ticking rate to the intensity of the light. This measurement is useful beyond the field of molecular clocks, as these nonlinear effects in optically trapped molecules may affect experiments ranging from cold chemistry studies to quantum computing platforms.

Despite some mysteries that remain to be understood, including the question of what process ultimately limits the coherence time of the molecular clock states, Kon's thesis work has delivered the promise of a working vibrational lattice clock that can be applied both to metrology in the terahertz frequency regime and to fundamental physics measurements. The future of Kon's experiment includes the enticing possibility of using the clock to compare the ticking rates of different molecular isotopologues. Combined with theory, this should allow not only unprecedented tests of quantum electrodynamics in heavy molecules, but also utilizing the clock as a quantum sensor to set limits on possible new forces, such as gravity-like interactions at the nanometer scale, that so far remain largely unexplored.

New York, NY, USA Tanya Zelevinsky

Acknowledgements

During my years at Columbia, I had the good opportunity to work with Tanya Zelevinsky, whose bold scientific vision, creativity, and keen intuition have influenced how I approach research. I heartily thank Tanya for having faith in me and cultivating a space for her students to fully mature into independent physicists. Tanya is a true auteur in the best sense. As a mentor, Tanya provided the invaluable advice of a seasoned scientist and kept our attention focused on the project milestones. The molecular lattice clock is Tanya's brilliant brainchild, and I'm proud to have contributed to bringing it to fruition.

If a scientist has seen further, it is by standing on the shoulders of giants. I owe a lot of my experimental skills to Stan Kondov, whose tremendous experience in the craft was instrumental in revitalizing a complex apparatus that was beginning to show its age. I'm also thankful to have closely shadowed Chih-Hsi Lee in my early years, who is boundlessly enthusiastic and has a comprehensive knowledge of the nuts and bolts of the experiment. In my first year, I was joined by Chris Liedl, who has a knack for finishing tasks most efficiently and a very admirable work ethic. I'm indebted to Hendrik Bekker, who brought a fresh scientific perspective and helped formulate the direction of my latter years. Most of all, I'd like to express my gratitude to Brandon Iritani, who is assiduous and has a solid intellectual drive. The most painstaking data-collecting sessions wouldn't be enjoyable if it weren't for his exuberant energy. Together, we succeeded in keeping up with the pace set by previous generations of the lab. I'm confident that his steadfastness would prove invaluable in achieving success in research and beyond. Special thanks also go to Jeff Sherman and his team at NIST for helping us get started with the GPS link.

Our experiments are joined at the hip with the theoretical efforts led by Robert Moszynski, whose wizardry has helped demystify our molecule. In particular, I thank Iwona Majewska for her readiness to perform numerous quantum chemistry calculations, all executed with incredible skill, precision, and rigor. Her theoretical findings are central to every aspect of Chih-Hsi's and my work, and I think of our theses as parts of a trilogy.

Journeying to new worlds is thrilling, but it takes time to acclimatize. Luckily, the latter can be softened by familiarity and kinship. I'd like to especially thank Aden

Lam for his constant friendship, inimitable intellect, humor, and broad culinary knowledge. Aden hailed from the same hometown and arrived at Columbia at the same time as I did. It's extraordinary to contemplate how much we have grown as scientists and people. I'm sure that he would continue to be first-rate wherever his adventures bring him. I also want to express my appreciation to physics department members for their camaraderie, including Debayan Mitra, Ivan Kozyryev, Konrad Wenz, Rees McNally, Geoff Iwata, Sebastian Vazquez-Carson, Jianhui Li, Jinyu Dai, Qi Sun, Varun Lochab and Aaron Liberman.

I wholeheartedly thank Jennifer Chiu, John Carr, Joey Cambareri, Michael Adan and Bob McGinnis for handling our equipment purchases, administrative tasks and infrastructural matters. I sincerely thank Michael Cabanlit and his team at Columbia Facilities for their promptness and proficiency. I tip my hat to Kevin and Diana for keeping our office and lab spaces clean. I also thank Kevin for keeping me on my toes with his thought-provoking physics and math questions, which never fail to brighten my day.

I'm very grateful to Sebastian Will and Ana Asenjo-Garcia for serving on my thesis committee and sharing their abundant wisdom. I warmly thank Jeremy Dodd and Josie Rosenfeld for lending the stage and special effects that gave me the outlet to develop my passion for teaching and science outreach. Back in school, I had no conception of higher education, much less a desire to pursue an academic career. I'm indebted to my teachers and mentors for taking me under their wing when I was but a Padawan, and for urging me to tread the path that ultimately brought me into this branch of physics: Rainer Dumke, Danny Segal, Richard Thompson, Florian Mintert, Rob Nyman, Tay Yian Ling, Ang Ghim Chee, Chan Hoong Leong, Chow Chi Wai and Liang Peiyun. Last but not least, I thank friends and family for their continuous support and companionship.

When I tasked myself with writing this section, I immediately remembered my time in New York City. My fondest memories in NYC tend to revolve around food. Among the numerous places that have kept me well-fed (which this margin is too narrow to contain), special shout-out goes to Living Thai, Miracle Thai, Bite of Thai, Momo Delight, Kalbi, Jan's Express, the Halal and breakfast cart on 116th Broadway, the breakfast (hot chocolate) cart on 114th Broadway, the helados cart on 110th Broadway (next to H Mart), the fruit stand outside of Five Guys on 110th Broadway, The Casbah on 66th street, Dun Huang, Grace Dodge Dining Hall, Appletree Market, Milano Market, Tom's Restaurant, Mill Korean, Wu and Nussbaum, Hungarian Pastry Shop, Hula Poke, Koronet Pizza, Famiglia, Westside Market, H Mart, Himalayan Curry House, Koko Wings, Banh, Naruto Ramen, Sushi Yasaka, Sushi W, Food Gallery 32, Nyonya, Noodle Village, and the annual Winter Village on Bryant Park. The American Museum of Natural History, where I learned that dinosaurs probably tasted like chicken, also holds a special place in my heart. Thanks to BookCulture and The Strand for keeping my love for written works alive, and AMC theatres at 84th street for many joyful movie nights. I'll always remember Central Park and its zoo on beautiful summer days. The vibrant colors and intensity of the swirling brushstrokes in Van Gogh's Starry Night can only be witnessed firsthand. It reminded me of how the night sky sparked my childhood curiosity about the natural world.

Work in this thesis benefited from the generous support of the Center for Fundamental Physics (CFP), the Templeton Foundation and the National Science Foundation (NSF).

Contents

Chapter 1
Introduction

1.1 Quantum Clocks in a Nutshell

The study of light and matter is intricately connected to the development of quantum metrology. Atomic and molecular spectra serve as ideal frequency standards. Every atom or molecule of the same element or composition is alike, and they exist in abundance throughout the universe as "cosmic pendulums"—according to our best understanding of physics, their unperturbed spectra should be Poincaré invariant. Quantum clocks were first seriously considered by Rabi, who, in his 1945 Richtmeyer Lecture, proposed the extension of his molecular beam resonance method to cesium atoms where microwaves tuned to a precise atomic frequency drive the transition of choice (also called the "clock transition"). Shortly after, in 1949, the National Bureau of Standards (now the National Institute of Standards and Technology) unveiled the world's first quantum (and molecular) clock with ammonia molecules based on its characteristic inversion transition. The abrogation of ephemeris time (based on Earth's orbit around the Sun) made way for the world's first atomic timescale when in 1967, the General Conference on Weights and Measures voted to define the SI second as "the duration of 9 192 631 770 periods of radiation" corresponding to the unperturbed ground state hyperfine splitting of ^{133}Cs; i.e., the unit of Hz (= 1-cycle per second implying 2π-radians per second) is such that this particular transition frequency is *exactly* 9 192 631 770 Hz. At the time of this writing, cesium fountain clocks realize the SI second with fractional accuracies in the low 10^{-16} and are the primary standards. The technical term "accuracy" is reserved for clocks that realize the standard second [1]; i.e., a clock whose unperturbed transition frequency in Hz has been measured against existing standards. When comparing the systematic frequency shifts of two clocks that outperform the current best standards, or whose clock frequencies are not yet known in Hz, the appropriate analogous term to accuracy is "systematic uncertainty".

Many techniques in modern atomic, molecular, and optical physics were invented to tackle the challenge of making a better quantum clock. This typically involves

K. H. Leung, *The Strontium Molecular Lattice Clock*, Springer Theses,
https://doi.org/10.1007/978-3-031-47647-1_1

gaining quantum control over the oscillator's external and internal degrees of freedom. Laser cooling, now the workhorse of many modern experiments to cool atoms and molecules to the μK-regime, greatly diminishes frequency shifts due to thermal motion. It is also usually the first step in creating ultracold gases with high phase-space densities. Equally important are techniques that manipulate or trap the atoms or molecules, such as the optical trapping of neutral particles. Particles held in a small region of space, with long-lived clock transitions, can be interrogated by electromagnetic radiation for long periods of time, thereby enabling finer frequency resolutions (i.e., narrower linewidths, δf) due to the Fourier uncertainty principle. Simultaneously interrogating a large number of particles reduces quantum projection noise (the fundamental variability in determining whether clock excitation has occurred, due to the stochastic nature of projective population measurements [2]), which for uncorrelated particles scales as $\sim 1/\sqrt{N}$ where N is the number of particles. "Instability" refers to the statistical variation of the measured clock frequencies, and is a measure of the level of precision accrued over a given averaging time. The fractional instability of a quantum clock can be written as[1]

$$\sigma_y(\tau) \sim \frac{\delta f}{f_{\text{clock}}\sqrt{N}} \frac{1}{\sqrt{\tau}} \sim \frac{1}{Q\sqrt{N\tau}}, \tag{1.1}$$

where τ is the averaging time (proportional to the number of successive measurements), f_{clock} is the clock frequency, and $Q \equiv f_{\text{clock}}/\delta f$ is the oscillator quality factor (Q-factor). Today's best oscillators are optical electronic transitions in atoms, with $Q > 10^{14}$. Typically $N \sim 10^3$ atoms are probed in a standing wave of laser light (called an optical lattice). In these optical lattice clocks, fractional instabilities below the 10^{-18} level have been demonstrated after several hours of averaging. Reliable clockwork and measurement of optical frequencies are enabled by femtosecond optical frequency combs, which provide a coherent link between optical and microwave frequencies.

1.2 Ultracold Molecules as Clocks and Sensors

By analogy with atomic lattice clocks, we can build *molecular* lattice clocks where the quantum oscillator involves the (de-)excitation of rotational or vibrational motion in the molecule or both. These rovibrational degrees of freedom are absent in atoms (where transitions involve either the electron or nuclear spin) and present a very rich physical system to study [Fig. 1.1]. Using lattice-clock techniques for

[1] Often in time and frequency literature, "y" denotes frequency-domain measurements (e.g., on a frequency counter), while "x" denotes time-domain measurements (e.g., on a time interval counter).

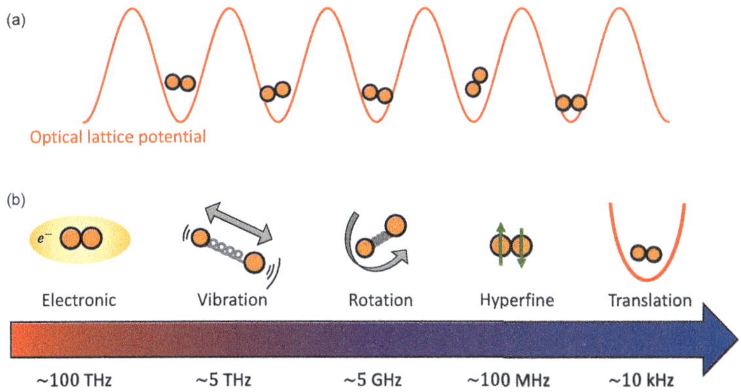

Fig. 1.1 (**a**) Cartoon illustration of a molecular lattice clock. Diatomic molecules (pairs of orange balls) at ultracold temperatures are held by an optical lattice (red curve) and probed with lasers along the lattice axis. (**b**) Hierarchy of transitions in a diatomic molecule and the typical transition frequencies. The rich internal structure opens up novel pathways in quantum science

molecular spectroscopy inherits many benefits of that architecture, such as the ability to resolve narrow molecular spectra and measure molecular energies with unprecedented accuracy.

From a purely metrological standpoint, the transition frequencies of rovibrational transitions are at least an order of magnitude smaller than optical electronic transitions. Equation (1.1) thus implies that a rovibrational molecular clock would have worse instability than an optical clock for the same length of interrogation time and particle numbers. Nevertheless, molecular clocks remain of interest within the frequency metrology community as they naturally realize THz frequency standards, filling the gap in the 0.1 THz–10 THz band where few secondary representations of the SI second currently exist. Equally important from a physicist's perspective, molecular clocks are attractive because molecular spectra offer enhanced sensitivities to several phenomena proposed by well-motivated extensions to the Standard Model.

For example, molecular spectra are intimately connected to the electron-to-proton mass ratio, $\eta \equiv m_e/m_p$, and the coupling of ultralight dark matter with ordinary matter may lead to a time variation in η [3–5]. Like a stretched spring, molecular vibrations arise due to the restoring force provided by the electronic molecular potential, $V(R)$, where R is the internuclear separation. Near the classical turning point, we can approximate $V(R)$ by a harmonic potential whose energy levels form a vibrational ladder with an angular frequency spacing of $\omega_e \sim \sqrt{\left(\frac{\partial^2 V}{\partial R^2}\right)/\mu_2} \propto \eta^{1/2}$, where μ_2 is the reduced mass of the dimer. Similarly, suppose we model a diatomic molecule as a rigid rotor with moment of inertia $\mu_2 R^2$, to first approximation. In that case, the levels in the rotational ladder will be separated by an angular frequency $\sim \hbar/(\mu_2 R^2) \propto \eta^1$. On the other hand, electronic transition frequencies depend solely on the fine-structure constant ($\alpha_{\rm fs}$).

By monitoring transitions with different sensitivities (or scaling) to η and α_{fs}, it becomes possible to detect changes in these fundamental constants over time, and constrain some classes of dark matter.

As another example, highly accurate molecular spectroscopy can set bounds on possible fifth forces at the length scale of a chemical bond (≈ 1Å) [6, 7]. Assuming hypothetical hadron-hadron interactions mediated by a new scalar boson between the nucleons of the two nuclei (having nucleon numbers A_1 and A_2), one may effectively model this with a Yukawa potential

$$V_5(R) = \hbar c \alpha_5 A_1 A_2 \frac{e^{-R/\lambda_5}}{R}, \tag{1.2}$$

where α_5 represents the coupling strength of the new interaction, and λ_5 is the Compton wavelength of the boson. The presence of $V_5(R)$ would shift the rovibrational energy levels proportionately with $A_1 A_2$. One promising way to constrain the size of α_5 is through the measurement of isotopologue shifts and making the comparison with quantum chemistry models.

This thesis describes experiments with a vibrational molecular lattice clock using ultracold diatomic strontium (Sr_2) molecules. Strontium has several properties well suited for the fundamental physics applications described above. As a group 2 (alkaline-earth metal) element in the periodic table, the spins of the two outermost electrons are oppositely paired (spin singlet) with zero orbital angular momentum in the ground state of the strontium atom, resulting in a *closed shell*. It follows that the molecular bonding of two strontium atoms in the ground state is mainly due to van der Waals forces, and the resulting ground potential in Sr_2 is a single isolated curve[2] that is amenable to modeling. Moreover, in a closed shell molecule, the energy levels in the ground potential are extremely insensitive to external static electric or magnetic fields, enabling accurate clock frequency measurements without meticulously characterizing stray background fields. A closely related property is that homonuclear molecules do not couple strongly to blackbody radiation, which might otherwise quench molecular state lifetimes or lead to large ac Stark frequency shifts due to the room-temperature environment. Dimer combinations consisting of strontium atoms with even nucleon numbers have no net nuclear spin[3] ($I = 0$). This implies the absence of hyperfine structure, greatly simplifying the observed spectra. While *ab initio* theoretical calculations of Sr_2 are not yet at the level of accuracy achieved for the simplest molecules (e.g., hydrogen isotopologues), they are comparatively more tractable than for dimers of alkaline-earth-like elements such as ytterbium (which has an unfilled f shell).

[2] Contrast this to bi-alkali metal molecules, e.g., KRb. Alkali metals have an unpaired valence electron. Thus, the combination of ground state K and Rb results in two distinct (singlet $X^1\Sigma^+$, or triplet $a^3\Sigma^+$) potentials.

[3] In fact, due to the Pauli exclusion principle, all even–even ground state nuclides have zero nuclear spin; i.e., even number of protons, and even number of neutrons.

There are also practical benefits from working with strontium. For example, optical lattice clocks based on strontium and other divalent atoms are poised to be primary standards when the SI second gets redefined in terms of an optical frequency. As such, there is a plethora of literature and technical expertise on matters ranging from measured physical properties, experimental techniques for quantum state control, and the production of ultracold gases of strontium and similar divalent atoms.

1.3 Molecular Quantum Science at the Dawn of the 2020s

Atoms and molecules are ubiquitous in nature, and studying their properties in a controlled laboratory environment improves our understanding of terrestrial and astrophysical chemical processes and, in some cases, can even reveal fundamental phenomena. Twentieth-century experiments with molecules emanating as beams represent a paradigm shift and formed the backdrop for the development of quantum mechanics [8]. Fruitful advancements from these early experiments include the discovery of spin angular momentum, radiation pressure from photon recoil (i.e., beam deflection), and nuclear magnetic resonance. Fast forward a century later, we are in the midst of yet another renaissance, where breakthroughs in photonics and laser technology enable experimentalists to exert control and precisely engineer the internal states of atoms and molecules for specialized tasks. Molecular quantum science has seen significant advancements over the past two decades, as documented in exhaustive review articles [9–14]. To add context and flavor to the presented research in this thesis, I give an abbreviated account of recent highlights that I found particularly noteworthy as a student in this field.

Research with cold ($\sim$ 1 mK) and ultracold ($\lesssim$ 1 μK) molecules, in broad strokes, fall into four categories: (1) precision tests of fundamental physics, (2) analog simulation of quantum many-body dynamics, (3) quantum information storage and processing, and (4) state-controlled cold chemistry. To meet the demands of these applications, precise preparation and detection of molecules in specialized states are needed, which in turn requires quantum control over multiple molecular degrees of freedom [Fig. 1.1b].

To date, two approaches have been successful in the creation of ultracold molecules [15]. The first, bottom-up approach starts with a trapped sample of laser-cooled constituent atoms that are subsequently associated into molecules. A remarkable example of this strategy is atom-by-atom assembly of single bi-alkali metal molecules using optical tweezers [16, 17]. While magnetoassociation[4] has been the de facto method for molecules with at least one magnetic atom (e.g., bi-

[4] Relying on the existence of an energetic avoided crossing between a bound molecular state and the scattering atoms in the vicinity of a Feshbach resonance (interspecies, if heteronuclear). An applied magnetic field is adiabatically swept to convert atoms to molecules, or vice versa.

alkalis [18]), the mobility of tightly-confining tweezer traps can also be leveraged to form molecules [19]. The former offers good prospects for mixed species with alkali and alkaline-earth-like metals constituents [20–22], while the latter may potentially find use for non-magnetic species such as Sr_2, in lieu of stimulated Raman passage [23] or photoassociation (see also Chaps. 2 and 4). Molecules of increasing complexity can be formed through such a modular approach, and indeed, magnetoassociation has been extended to produce an ultracold gas of triatomic molecules [24, 25]. Because the laser cooling of atoms is very robust [26], the resulting molecules often inherit high phase-space densities [27]. Inelastic collisions between molecules can lead to population loss (see also Chap. 4). These undesirable losses can be circumvented by engineering collisional shielding using microwave-dressing [28–30] or dipolar interactions in restricted geometries [31]. By doing so, the field has achieved the long-standing goal of cooling a dipolar molecular gas into quantum degeneracy. The second, top-down approach begins with a cryogenic buffer gas beam of slow-moving molecules directly cooled with lasers (or other optoelectric methods) [32]. Efficient photon cycling requires the molecular bond length to remain essentially unchanged upon electronic excitation,[5] restricting the molecule class. Nevertheless, (magneto-)optical trapping of neutral molecules has been successfully realized for alkaline-earth metal monofluorides [33, 34], a lanthanide-like metal monooxide [35], and even a *polyatomic* alkaline-earth metal monohydroxide [36].

Long trapping lifetimes afforded by conservative potentials lay the bedrock for coherent state manipulation. Hyperfine transitions are particularly suited for maintaining quantum superpositions over long periods of time—nuclear spin flips are impervious to external perturbations. Nuclear coherence times at the few-second scale have been shown for several bi-alkali-metal molecular species [37–39]. We can enrich the scope of physical phenomena when multiple molecules interact. In this regard, the rotational levels of polar molecules are a natural resource because spin-exchange can be engineered via strong dipole-dipole interactions, allowing for the deterministic creation of Bell pairs (where two molecules are maximally entangled) [40, 41]. To this end, it is advantageous for the rotational coherence time to be much longer than the duration for spin-exchange, motivating the study of magic traps [42–44] (see also Chap. 5). Interatomic Rydberg interactions have become a vital tool for atom-based quantum processing platforms [45, 46]. Quantum gas microscopy has been employed to observe the formation of Rydberg macrodimers in optical lattices [47]. These micron-sized "artificial" molecules have bond lengths that are nearly 3 orders of magnitude greater than that of $^4He_2$, the largest naturally occurring ground state dimer. The mechanism of Rydberg blockade has been recently observed between an atom and a *molecule* [48], suggesting the intriguing prospect of utilizing

[5] In the case of $^{88}Sr_2$, although rotational closure can be guaranteed for $J = 2 \leftrightarrow J' = 2$ transitions (due to the absence of $J = 1$ in the ground potential), diagonal Franck-Condon factors are not present for transitions from the ground potential to the low-lying excited potentials to the best of our knowledge.

neighboring atoms as information buses for transducing and reading out (possibly non-destructively) the corresponding molecular state with high fidelity. The use of an information bus hearkens back to quantum logic spectroscopy [49], which is a versatile technique used in trapped ion systems. By prepending the spectroscopy with a quantum logic sequence that projects the ion in a pure rotational state, THz-rotational transitions in CaH^+ were directly driven using a frequency comb [50, 51], achieving an impressive precision at the 10^{-13} level [52]. As cotrapped ions share motional modes, a series of species-targeted motional sideband transitions in such a hybrid platform can be used to entangle the internal states of the molecular and atomic ions [53].

High-precision laser spectroscopy of molecular vibrations has been demonstrated in diverse experimental settings. In homonuclear molecules (including the isotopologue of Sr_2 studied in this thesis), within the same nuclear spin multiplicity, spin-statistics and angular momentum selection rules dictate that only electric quadrupole single-photon transitions are allowed between vibrational states of the ground potential. In neutral H_2, the large circulating optical powers in a high-finesse cavity facilitated the first Lamb-dip saturation spectroscopy of an electric quadrupole transition in the (2-0) band at a fractional resolution of 3×10^{-11} [54]. In the simplest heteronuclear molecule HD^+, an ion, laser excitation of the fundamental vibrational overtone transition in the ground potential was observed as early as the 1970s by Lamb [55], who later suggested that the rovibrational spectra of such simple molecules might serve as fertile testing grounds for few-body quantum electrodynamics. Global analyses of high precision HD^+ spectroscopy have put to test the foundational aspects of quantum theory and beyond Standard Model forces [7, 56]. Recent experiments with trapped HD^+ ions probed a near-infrared single-photon transition starting from the rovibrational ground state, achieving a Q-factor of 6×10^{11} and a fractional uncertainty of 2.3×10^{-12} for its nuclear spin-averaged transition frequency [56, 57]. Owing to the calculability of HD^+ spectra from first principles, the absolute value of the electron-to-proton mass ratio, η, can be extracted from comparisons between theory and experimental data. Through this approach, rovibrational spectroscopy of HD^+ have obtained η at a competitive precision compared to other contemporary methods [58–60]. Networks of atomic clocks are already setting the most stringent bounds on the temporal stability of the fundamental constants [61–63]. Diversifying these networks to include molecular clocks, whose rovibrational "ticking" mechanism is different from atomic clocks, should augment efforts in this direction. Microwave spectroscopy of near degenerate transitions in the vibrational ladders of the $X^1\Sigma^+$ and $a^3\Sigma^+$ potentials in KRb is a notable example of a molecule-based experiment constraining the time-variation of η [64].

Cold molecule "tabletop" experiments are setting ever tighter limits on the electron's permanent electric dipole moment (eEDM) [65, 66]. A non-zero eEDM would signify a violation of time-reversal symmetry, with implications for the preponderance of matter over antimatter in the observable universe and the existence of beyond Standard Model (e.g., supersymmetric) particles [67]. So far, all experiments report an eEDM consistent with zero—the current record [65] constrains its value to be $\lesssim 4 \times 10^{-30}\,e\,\text{cm}$, a thousand-fold improvement from the first molecule-based measurement just a decade ago [68, 69]. Typically, an external electric field is applied to polarize the molecule in the laboratory frame and make the eEDM measurable. An advantage of polar molecules over atoms is that they possess closely spaced Ω-doublets of opposite parity, requiring orders of magnitude smaller field strengths for full polarization. The eEDM signal is extracted from the differential shifts of the Ω-doublets—this is effectively equivalent to reversing the orientation of the electric field experienced by the electron relative to its spin, mitigating a key systematic. Such precision measurement experiments would undoubtedly benefit from the arsenal of tricks that are becoming available for molecular quantum control.

1.4 Outline of this Thesis

- In Chap. 2, I briefly review the structure of atomic and diatomic strontium and outline the experimental apparatus for producing ultracold $^{88}Sr_2$ molecules via narrow-line photoassociation and the detection scheme based on photofragmentation. A theoretical treatment of trapped particles in a one-dimensional optical lattice and relevant experimental techniques such as carrier thermometry and resolved sideband spectroscopy are given.
- In Chap. 3, I present up-to-date results on our spectroscopy of the electronic ground potential $X^1\Sigma_g^+$, and the singly excited potentials $(1)0_u^+$ and $(1)1_u$ that asymptote to the intercombination line. Next, I describe the stabilization of the metrological grade master laser to a high finesse optical cavity and the working principle of an optical frequency comb. Finally, I construct a model of light-molecule interactions based on the optical Bloch equations, supplementing the discussion with experimental observations of phenomena such as two-photon Rabi oscillations, Autler-Townes doublets, electromagnetically induced transparency, etc.
- In Chap. 4, I report the efficient creation of an ultracold gas of strontium molecules in the absolute rovibrational ground state, accomplished using stimulated Raman adiabatic passage. Transition strength measurements of $X^1\Sigma_g^+$ to $(1)0_u^+$ lines are presented. I further describe the exploration of two-body collisions with absolute ground state molecules.
- In Chap. 5, I report a vibrational molecular clock with 4.6×10^{-14} systematic uncertainty. I begin by outlining the magic wavelength protocol, discussing factors that limit the molecular lifetime, and making the connection between polarizability and transition strength. Transition strength measurements of $X^1\Sigma_g^+$ to $(1)1_u$ lines are presented. Finally, I describe the full systematic evaluation of the molecular clock and the measurement of its absolute frequency with 13-digit precision via the GPS satellite constellation.

Parts of this thesis have been published in Refs. [70–73].

References

1. Marlow, B.L.S., Scherer, D.R.: A review of commercial and emerging atomic frequency standards. IEEE Trans. Ultrason. Ferroelectr. Freq. Control **68**, 2007–2022 (2021)
2. Itano, W.M., et al.: Quantum projection noise: population fluctuations in two-level systems. Phys. Rev. A **47**, 3554 (1993)
3. Zelevinsky, T., Kotochigova, S., Ye, J.: Precision test of mass-ratio variations with lattice-confined ultracold molecules. Phys. Rev. Lett. **100**, 043201 (2008)
4. Oswald, R., et al.: Search for dark-matter-induced oscillations of fundamental constants using molecular spectroscopy. Phys. Rev. Lett. **129**, 031302 (2022)
5. Kozyryev, I., Lasner, Z., Doyle, J.M.: Enhanced sensitivity to ultralight bosonic dark matter in the spectra of the linear radical SrOH. Phys. Rev. A **103**, 043313 (2021)
6. Salumbides, E.J., et al.: Bounds on fifth forces from precision measurements on molecules. Phys. Rev. D **87**, 112008 (2013)
7. Germann, M., et al.: Three-body QED test and fifth-force constraint from vibrations and rotations of HD^+. Phys. Rev. Res. **3**, L022028 (2021)
8. Ramsey, N.F.: Molecular beams: our legacy from Otto Stern. Z. Phys. D: At. Mol. Clusters **10**, 121–125 (1988)
9. Langen, T., Valtolina, G., Wang, D., Ye, J.: Quantum state manipulation and science of ultracold molecules. arXiv preprint arXiv:2305.13445 (2023)
10. Softley, T.P.: Cold and ultracold molecules in the twenties. Proc. R. Soc. Lond. A **479**, 20220806 (2023)
11. Bohn, J.L., Rey, A.M., Ye, J.: Cold molecules: progress in quantum engineering of chemistry and quantum matter. Science **357**, 1002–1010 (2017)
12. Carr, L.D., DeMille, D., Krems, R.V., Ye, J.: Cold and ultracold molecules: science, technology and applications. New J. Phys. **11**, 055049 (2009)
13. Bell, M.T., Softley, T.: Ultracold molecules and ultracold chemistry. Mol. Phys. **107**, 99–132 (2009)
14. Krems, R., Friedrich, B., Stwalley, W.C.: Cold Molecules: Theory, Experiment, Applications. CRC Press, Boca Raton (2009)
15. Heazlewood, B.R., Softley, T.P.: Towards chemistry at absolute zero. Nat. Rev. Chem. **5**, 125–140 (2021)
16. Yu, Y., et al.: Coherent optical creation of a single molecule. Phys. Rev. X **11**, 031061 (2021)
17. Kaufman, A.M., Ni, K.-K.: Quantum science with optical tweezer arrays of ultracold atoms and molecules. Nat. Phys. **17**, 1324–1333 (2021)
18. Zhang, J.T., et al.: Forming a single molecule by magnetoassociation in an optical tweezer. Phys. Rev. Lett. **124**, 253401 (2020)
19. Ruttley, D.K., et al.: Formation of ultracold molecules by merging optical tweezers. Phys. Rev. Lett. **130**, 223401 (2023)
20. Barbé, V., et al.: Observation of Feshbach resonances between alkali and closed-shell atoms. Nat. Phys. **14**, 881–884 (2018)
21. Green, A., et al.: Feshbach resonances in p-wave three-body recombination within fermi-fermi mixtures of open-shell ^6Li and closed-shell ^{173}Yb atoms. Phys. Rev. X **10**, 031037 (2020)
22. Franzen, T., et al.: Observation of magnetic Feshbach resonances between Cs and ^{173}Yb. Phys. Rev. Res. **4**, 043072 (2022)
23. Stellmer, S., Pasquiou, B., Grimm, R., Schreck, F.: Creation of ultracold Sr_2 molecules in the electronic ground state. Phys. Rev. Lett. **109**, 115302 (2012)
24. Yang, H., et al.: Creation of an ultracold gas of triatomic molecules from an atom–diatomic molecule mixture. Science **378**, 1009–1013 (2022)
25. Yang, H., et al.: Evidence for the association of triatomic molecules in ultracold $^{23}Na^{40}K+\ ^{40}K$ mixtures. Nature **602**, 229–233 (2022)
26. Schreck, F., Druten, K.v.: Laser cooling for quantum gases. Nat. Phys. **17**, 1296–1304 (2021)

27. De Marco, L., et al.: A degenerate Fermi gas of polar molecules. Science **363**, 853–856 (2019)
28. Schindewolf, A., et al.: Evaporation of microwave-shielded polar molecules to quantum degeneracy. Nature **607**, 677–681 (2022)
29. Anderegg, L., et al.: Observation of microwave shielding of ultracold molecules. Science **373**, 779–782 (2021)
30. Karman, T., Hutson, J.M.: Microwave shielding of ultracold polar molecules. Phys. Rev. Lett. **121**, 163401 (2018)
31. Matsuda, K., et al.: Resonant collisional shielding of reactive molecules using electric fields. Science **370**, 1324–1327 (2020)
32. Fitch, N.J., Tarbutt, M.R.: Laser-cooled molecules. Adv. At. Mol. Opt. Phys. **70**, 157–262 (2021)
33. Anderegg, L., et al.: An optical tweezer array of ultracold molecules. Science **365**, 1156–1158 (2019)
34. Langin, T.K., Jorapur, V., Zhu, Y., Wang, Q., DeMille, D.: Polarization enhanced deep optical dipole trapping of Λ-cooled polar molecules. Phys. Rev. Lett. **127**, 163201 (2021)
35. Wu, Y., Burau, J.J., Mehling, K., Ye, J., Ding, S.: High phase-space density of laser-cooled molecules in an optical lattice. Phys. Rev. Lett. **127**, 263201 (2021)
36. Vilas, N.B., et al.: Magneto-optical trapping and sub-Doppler cooling of a polyatomic molecule. Nature **606**, 70–74 (2022)
37. Park, J.W., Yan, Z.Z., Loh, H., Will, S.A., Zwierlein, M.W.: Second-scale nuclear spin coherence time of ultracold ^{23}Na^{40}K molecules. Science **357**, 372–375 (2017)
38. Gregory, P.D., Blackmore, J.A., Bromley, S.L., Hutson, J.M., Cornish, S.L.: Robust storage qubits in ultracold polar molecules. Nat. Phys. **17**, 1149–1153 (2021)
39. Lin, J., He, J., Jin, M., Chen, G., Wang, D.: Seconds-scale coherence on nuclear spin transitions of ultracold polar molecules in 3D optical lattices. Phys. Rev. Lett. **128**, 223201 (2022)
40. Bao, Y., et al.: Dipolar spin-exchange and entanglement between molecules in an optical tweezer array. arXiv preprint arXiv:2211.09780 (2022)
41. Holland, C.M., Lu, Y., Cheuk, L.W.: On-demand entanglement of molecules in a reconfigurable optical tweezer array. arXiv preprint arXiv:2210.06309 (2022)
42. Gregory, P.D., et al.: Second-scale rotational coherence and dipolar interactions in a gas of ultracold polar molecules. arXiv preprint arXiv:2306.02991 (2023)
43. Park, A.J., et al.: Extended rotational coherence of polar molecules in an elliptically polarized trap. arXiv preprint arXiv:2306.07264 (2023)
44. Bause, R., et al.: Tune-out and magic wavelengths for ground-state ^{23}Na ^{40}K molecules. Phys. Rev. Lett. **125**, 023201 (2020)
45. Evered, S.J., et al.: High-fidelity parallel entangling gates on a neutral atom quantum computer. arXiv preprint arXiv:2304.05420 (2023)
46. Scholl, P., et al.: Erasure conversion in a high-fidelity Rydberg quantum simulator. arXiv preprint arXiv:2305.03406 (2023)
47. Hollerith, S., et al.: Quantum gas microscopy of Rydberg macrodimers. Science **364**, 664–667 (2019)
48. Guttridge, A., et al.: Observation of Rydberg blockade due to the charge-dipole interaction between an atom and a polar molecule. Phys. Rev. Lett. **131**, 013401 (2023)
49. Wolf, F., et al.: Non-destructive state detection for quantum logic spectroscopy of molecular ions. Nature **530**, 457–460 (2016)
50. Chou, C.-w., et al.: Preparation and coherent manipulation of pure quantum states of a single molecular ion. Nature **545**, 203–207 (2017)
51. Chou, C.-w., et al.: Frequency-comb spectroscopy on pure quantum states of a single molecular ion. Science **367**, 1458–1461 (2020)
52. Collopy, A.L., Schmidt, J., Leibfried, D., Leibrandt, D.R., Chou, C.-W.: Effects of an oscillating electric field on and dipole moment measurement of a single molecular ion. Phys. Rev. Lett. **130**, 223201 (2023)

53. Lin, Y., Leibrandt, D.R., Leibfried, D., Chou, C.-w.: Quantum entanglement between an atom and a molecule. Nature **581**, 273–277 (2020)
54. Cozijn, F.M.J., Diouf, M.L., Ubachs, W.: Lamb dip of a quadrupole transition in H_2. Phys. Rev. Lett. **131**, 073001 (2023)
55. Wing, W.H., Ruff, G.A., Lamb Jr., W.E., Spezeski, J.J.: Observation of the infrared spectrum of the hydrogen molecular ion HD^+. Phys. Rev. Lett. **36**, 1488 (1976)
56. Alighanbari, S., Kortunov, I., Giri, G., Schiller, S.: Test of charged baryon interaction with high-resolution vibrational spectroscopy of molecular hydrogen ions. Nat. Phys. **19**, 1263–1269 (2023)
57. Alighanbari, S., Hansen, M.G., Korobov, V., Schiller, S.: Rotational spectroscopy of cold and trapped molecular ions in the Lamb–Dicke regime. Nat. Phys. **14**, 555–559 (2018)
58. Kortunov, I., et al.: Proton–electron mass ratio by high-resolution optical spectroscopy of ion ensembles in the resolved-carrier regime. Nat. Phys. **17**, 569–573 (2021)
59. Patra, S., et al.: Proton-electron mass ratio from laser spectroscopy of HD^+ at the part-per-trillion level. Science **369**, 1238–1241 (2020)
60. Alighanbari, S., Giri, G., Constantin, F.L., Korobov, V., Schiller, S.: Precise test of quantum electrodynamics and determination of fundamental constants with HD^+ ions. Nature **581**, 152–158 (2020)
61. Lange, R., et al.: Improved limits for violations of local position invariance from atomic clock comparisons. Phys. Rev. Lett. **126**, 011102 (2021)
62. Schwarz, R., et al.: Long term measurement of the ^{87}Sr clock frequency at the limit of primary Cs clocks. Phys. Rev. Res. **2**, 033242 (2020)
63. Kobayashi, T., et al.: Search for ultralight dark matter from long-term frequency comparisons of optical and microwave atomic clocks. Phys. Rev. Lett. **129**, 241301 (2022)
64. Kobayashi, J., Ogino, A., Inouye, S.: Measurement of the variation of electron-to-proton mass ratio using ultracold molecules produced from laser-cooled atoms. Nat. Commun. **10**, 3771 (2019)
65. Roussy, T.S., et al.: An improved bound on the electron's electric dipole moment. Science **381**, 46–50 (2023)
66. ACME Collaboration: Andreev, V., Ang, D.G., DeMille, D., Doyle, J.M., Gabrielse, G., Haefner, J., Hutzler, N.R., Lasner, Z., Meisenhelder, C., O'Leary, B.R., Panda, C.D., West, A.D., West, E.P., Wu, X.: Improved limit on the electric dipole moment of the electron. Nature **562**, 355–360 (2018)
67. Cairncross, W.B., Ye, J.: Atoms and molecules in the search for time-reversal symmetry violation. Nat. Rev. Phys. **1**, 510–521 (2019)
68. Hudson, J.J., et al.: Improved measurement of the shape of the electron. Nature **473**, 493–496 (2011)
69. Moskowitz, C.: Electron appears spherical, squashing hopes for new physics theories. Sci. Am. **11** (2013). https://www.scientificamerican.com/article/electron-spherical-electric-dipole-moment/
70. Leung, K.H., et al.: Transition strength measurements to guide magic wavelength selection in optically trapped molecules. Phys. Rev. Lett. **125**, 153001 (2020)
71. Mitra, D., Leung, K.H., Zelevinsky, T.: Quantum control of molecules for fundamental physics. Phys. Rev. A **105**, 040101 (2022)
72. Leung, K.H., et al.: Ultracold $^{88}Sr_2$ molecules in the absolute ground state. New J. Phys. **23**, 115002 (2021)
73. Leung, K.H., et al.: Terahertz vibrational molecular clock with systematic uncertainty at the 10^{-14} level. Phys. Rev. X **13**, 011047 (2023)

Chapter 2
Molecular Structure and Production of Ultracold $^{88}Sr_2$ in an Optical Lattice

2.1 ^{88}Sr Atomic Structure and Laser Cooling

A complete account of the experimental apparatus can be found in Ref. [1]. Here, I give a summary of the cooling sequence. The relevant level structure of ^{88}Sr (an alkaline-earth metal) is illustrated by Fig. 2.1a, showing characteristic singlet and triplet states due to having two valence electrons in its outermost electronic shell. ^{88}Sr also has zero nuclear spin ($I = 0$), and therefore the levels have no further substructure. The electronic ground state has the term symbol $^1S_0$; i.e., total electronic spin $s = 0$, total electronic angular momentum $l = 0$, resulting in total electronic angular momentum $j = 0$. Comprehensive tables of atomic strontium transitions and their transition strengths can be found in Ref. [2].

Singlet-singlet transitions are the strongest, and the $^1S_0 \leftrightarrow {}^1P_1$ transition at 461 nm is used to decelerate a beam of hot strontium atoms emanating from an oven heated to above 450 °C, Zeeman slowing the atoms to be within the capture velocity of the first-stage "blue" magneto-optical trap (MOT) that also operates on the strong 461 nm transition. A three-dimensional (3D) MOT uses a spatially varying quadrupole magnetic field and six circularly polarized counter-propagating beams along all three axes to generate the restoring and dissipative forces, as illustrated by Fig. 2.1b and c. This cools the atoms close to the Doppler limit, achieving temperatures $\sim \hbar\Gamma_a/(2k_B)$, where Γ_a is the natural linewidth of the cooling transition. For the 461 nm transition, $\Gamma_a \approx 2\pi \times 32$ MHz resulting in a temperature of $\sim$ 800 μK. This is still much hotter than the typical depths of an optical trap with today's technology. For divalent atoms such as strontium, it is a common strategy to implement a second-stage "red" MOT, which cools using the intercombination line $^1S_0 \leftrightarrow {}^3P_1$ at 689 nm. This is a singlet-triplet transition involving the spin flip of an electron ($s = 0 \rightarrow s' = 1$) that is allowed due to spin-

K. H. Leung, *The Strontium Molecular Lattice Clock*, Springer Theses,
https://doi.org/10.1007/978-3-031-47647-1_2

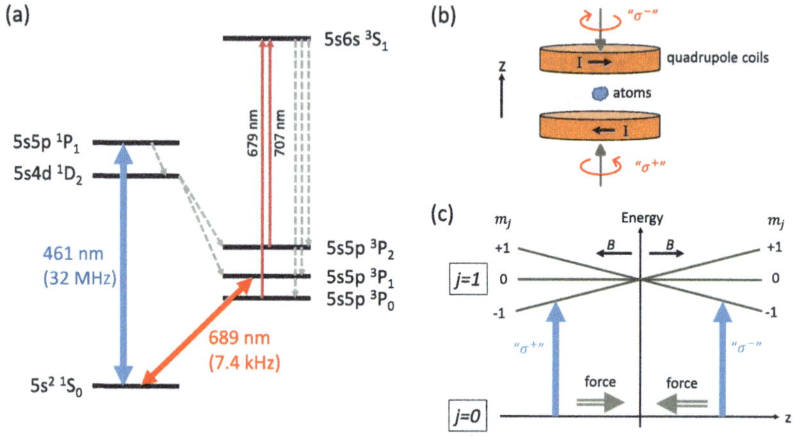

Fig. 2.1 (**a**) Simplified strontium atom level diagram (not to scale), showing the relevant transitions for laser cooling (461 nm blue and 689 nm red). Bracketed values are the respective natural linewidths. Gray arrows indicate decay channels plugged with the repumpers (679 nm and 707 nm). (**b**) Cartoon illustration of a magneto-optical trap (MOT). High current coils in the anti-Helmholtz configuration (currents flow in opposite directions) produce a spatially varying quadrupole B-field. Circularly polarized cooling light (i.e., MOT beam) propagates along the B-field gradient, through the coils, and addresses the atoms (shown only for the $\hat{Z}$-axis). Due to the sign flip in the B-field direction at the origin, the polarization of the MOT beams is typically a source of confusion for the experimentalist—technically, the MOT beams should drive σ^- transitions if we always take the quantization axis to be along the local direction of the B-field. The unambiguous method (a literal rule of thumb) to determine required circular polarization is to align one's *left* hand thumb along the local direction of the B-field. The polarization vector of the laser beam providing the restoring optical force should rotate along the curl of the fingers on the same left hand. When using a commercial polarimeter, care should also be taken to check the equipment's handedness convention (i.e., from the point of view of the source versus the receiver). (**c**) Working principle of a type-I MOT for the $S \leftrightarrow P$ cooling transitions in strontium. Atoms away from the center of the MOT experience both restoring and Doppler cooling forces when the laser light is slightly red-detuned from the transition by typically a few Γ_a

orbit coupling.[1] Thus, it has a narrower linewidth at $\Gamma_a \approx 2\pi \times 7.4$ kHz implying a Doppler temperature of $\sim 0.2\ \mu$K. Under realistic conditions, we normally achieve atom temperatures of about $\sim 2\ \mu$K, inferred from the time-of-flight expansion[2] of the atomic gas. Typical atom numbers captured in the red MOT are $\sim 10^6$, which may improve by refilling the strontium source in the oven and declogging the nozzle. At this point, the atoms are cold enough to be loaded into the optical lattice trap and photoassociated into molecules.

[1] The valence electrons are moving close to relativistic speeds due to the large proton number in strontium.

[2] The Gaussian width of a freely expanding thermal cloud evolves as $\sigma_j = \sqrt{\sigma_{0,j}^2 + t^2 k_B T_{\mathrm{Sr},j}/m_{\mathrm{Sr}}}$. For short times-of-flight, the cloud still "remembers" the shape of its trap and $k_B T_{\mathrm{Sr},j} = m_{\mathrm{Sr}}\sigma_j^2/(1/\omega_{\mathrm{trap},j}^2 + t^2)$, where $\omega_{\mathrm{trap},j}$ and $T_{\mathrm{Sr},j}$ are the angular trap frequency and temperature of the atomic cloud in the j direction, and t is the time-of-flight.

2.2 $^{88}Sr_2$ Molecular Structure and Selection Rules

2.2.1 Molecular Term Symbols

For diatomic molecules with strong spin-orbit coupling, the total electronic orbital $(\mathbf{l}_1 + \mathbf{l}_2)$ and spin $(\mathbf{s}_1 + \mathbf{s}_2)$ angular momenta of the composite two-atom system are coupled and not conserved. Thus, the eigenvalues Λ and Σ of the **projections** of $(\mathbf{l}_1 + \mathbf{l}_2)$ and $(\mathbf{s}_1 + \mathbf{s}_2)$ onto the internuclear axis respectively are *not* good quantum numbers. However, the sum $\Omega = \Lambda + \Sigma$ is a good quantum number. $\mathbf{J}$ is the operator representing the **total** angular momentum of the molecule in the space-fixed (or lab) frame; i.e., the sum of the rotation of the nuclei (about their center of mass and an axis perpendicular to the internuclear axis) *and* the total angular momentum of the electrons. The eigenvalue of $(\mathbf{J} \cdot \mathbf{J})$ is $J(J+1)$, and the eigenvalue of the projection of $\mathbf{J}$ onto the space-fixed quantization axis is M. Since the **total** angular momentum is conserved, J and M are good quantum numbers. Note that Ω can also be thought of as the eigenvalue of the projection of $\mathbf{J}$ onto the internuclear axis in the body-fixed (or molecule) frame; i.e., in this frame, the molecule is not rotating, and the total angular momentum comes solely from the electrons. Putting it all together, an angular momentum state of the dimer is fully specified by $|J, M, \Omega\rangle$.

To categorize the different situations that arise when angular momenta in a molecule couple together, physicists and chemists use **Hund's cases** which describe idealized limiting cases. Actual molecules often do not fall strictly into a particular Hund's case. Note that for each J, the size of the Hilbert state space is the same regardless of the choice of Hund's case basis sets [3, 4].

Molecular term symbols in Hund's case (c) are written as

$$|\Omega|^{+/-}_{g/u} .$$

The superscript "$+/-$" denotes the reflection symmetry of the electronic wavefunction about a plane containing the internuclear axis and is only written for $\Omega = 0$ states. The subscript "g/u" denotes *gerade* (German for even) or *ungerade* (German for odd) symmetry from the inversion of the electronic wavefunction through the center point of symmetry and is only written for homonuclear molecules (identical nuclei).

Molecular term symbols in Hund's case (a) are written as

$$^{2S+1}|\Lambda|^{+/-}_{g/u} ,$$

where S is the total electronic spin quantum number.[3] The superscript "$+/-$" is only written for $\Lambda = 0$. Analogous to atomic term symbols, $|\Lambda| = 0, 1, 2, \ldots$ are written as $\Sigma, \Pi, \Delta, \ldots$.

Terms symbols are prefixed by either letters or bracketed numbers, with "X" being reserved for the absolute ground potential. Letters are capitalized if the

[3] The eigenvalue of $|\mathbf{s}_1 + \mathbf{s}_2|^2$ is $S(S+1)$.

excited potential has the same spin multiplicity as X, and non-capitalized if otherwise. Empirically, the letters ascend alphabetically in order of increasing energy, e.g., $A, B, C, \ldots$, but older literature may not follow this convention. Bracketed numbers, e.g., (1), (2), ..., usually clarify this ambiguity.

2.2.2 *Ground Electronic Potential asymptoting to* $^1S_0 + {}^1S_0$

The ground electronic potential of Sr_2 asymptotes to the $^1S_0 + {}^1S_0$ dissociation threshold, and has the term symbol $X0_g^+$ in case (c) notation. It is easy to see why: $\Omega = 0$ because l_1, l_2, s_1, s_2 of the constituent atoms are zero so there is no electronic angular momentum to project, and the reflection and inversion symmetries must be "+" and gerade as the atoms are in identical electronic states. The generalization of arguments like these to arrive at the possible molecular term symbol(s) given those of the constituent atoms is called the **Wigner-Witmer correlation rules**. Even though the ground state has $\Lambda = 0$ and $S = 0$, it is not uncommon to see the potential labeled with the case (a) term symbol $X^1\Sigma_g^+$ in literature. Figure 2.2a shows the empirical $X^1\Sigma_g^+$ potential from Ref. [5] obtained via Fourier-transform spectroscopy.

Here, the total electronic angular momentum is zero. In particular, the net electronic spin is also zero (i.e., for X states of Sr_2, J is also the rotational quantum number[4] N), so $\mathbf{J}$ is purely due to the relative rotation of the nuclei. For homonuclear dimers with bosonic nuclei in a symmetric nuclear spin configuration (e.g., $^{88}Sr_2$, where the total nuclear spin is zero), spin-statistics forces J to be even for 0_g^+ states, and odd for 0_u^+ states. The argument is as follows. The effect of exchanging identical nuclei on the spatial wavefunction of the dimer is equivalent to performing a parity transformation on the entire molecular rotational wavefunction (which includes the nuclei), followed by the inversion of the electronic wavefunction. The rotational wavefunctions of a symmetric top[5] are Wigner D-functions; i.e., $\langle \phi, \theta, \chi | J, M, \Omega \rangle = \sqrt{\frac{2J+1}{8\pi^2}} D^{J*}_{M\Omega}(\phi, \theta, \chi)$, here $\{\phi, \theta, \chi\}$ are the Euler angles that bring the space-fixed frame into alignment with the body-fixed frame [6]. For $\Omega = 0$, the Wigner D-functions reduce to the spherical harmonics with parity eigenvalues $(-1)^J$. By definition, inversion of the electronic wavefunction gives a factor of $(+1)$ for gerade states and (-1) for ungerade states. But the overall product must be $(+1)$ since the nuclei are bosons and in a symmetric nuclear spin configuration.[6] Therefore J must be even in 0_g^+, and odd in 0_u^+.

---

[4] Following conventional literature, the difference between N and J is that N omits the electronic spin. Note that the nuclear spin is omitted in both N and J.

[5] A dimer is technically a trivial case of a symmetric top.

[6] Recall that the rules pertaining to the permutation symmetry of indistinguishable quantum particles apply to the *overall* quantum state of the ensemble; i.e., the tensor product of the spatial and spin states.

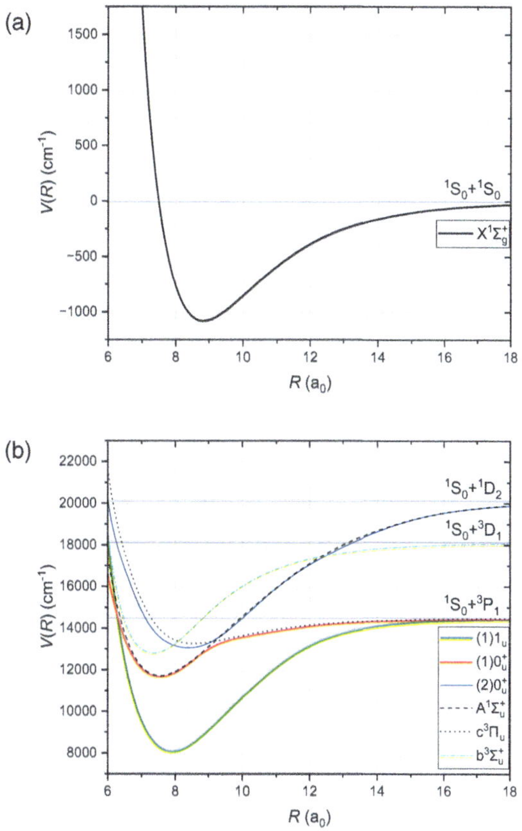

Fig. 2.2 Molecular potentials of Sr_2. Gray solid horizontal lines indicate the atomic dissociation thresholds as labeled. (**a**) Electronic ground potential $X^1\Sigma_g^+$ (or $X0_g^+$) correlating to the $^1S_0 + {}^1S_0$ threshold (empirical curve from Ref. [5]). (**b**) A small selection of ungerade excited electronic potentials. The mixing of non-relativistic potentials (given in Hund's case (**a**) term symbols, and taken from Refs. [7, 8]) through the spin-orbit interaction results in the relativistic potentials $(1)0_u^+$ and $(1)1_u$ (empirical curves from Ref. [9]). Plotted are potentials that extend below the $^1S_0 + {}^3P_1$ threshold. Not all excited potentials are shown to avoid clutter (comprehensive plots of potentials up to the $^1S_0 + {}^1P_1$ threshold omitted here can be found in Refs. [10, 11])

This implies that J **must be even in the $X^1\Sigma_g^+$ ground potential** of the bosonic dimer. All 63 bound vibrational states with $J = 0, 2$ in $^{88}Sr_2$ have been found in our experiment, and their energies are reported in Sect. 3.6. The fact that odd J are absent in our ground state spectra can be considered as a verification of the spin-statistics theorem.

2.2.3 *Excited Electronic Potentials asymptoting to $^1S_0 + {}^3P_1$*

From the Wigner-Witmer rules, the possible molecular states that can arise from the combination of an atom in $^1S_0$ with another atom in $^3P_1$ are $^3\Sigma^+$ and $^3\Pi$, with no restriction on the inversion symmetry. This results in a total of four potentials that asymptote to $^1S_0 + {}^3P_1$, namely $a^3\Sigma_u^+$, $c^3\Pi_u$, $^3\Sigma_g^+$ and $^3\Pi_g$. Just as for atomic Sr, the heavy mass of the nucleus leads to strong **spin-orbit mixing** of the excited singlet and triplet rovibronic states in molecular Sr_2. To properly account for experimental observations, Skomorowski et al. [7] found it necessary to admix the ungerade $a^3\Sigma_u^+$, $c^3\Pi_u$ with the higher lying ungerade potentials $A^1\Sigma_u^+$, $b^3\Sigma_u^+$, $B^1\Sigma_u^+$, and $B'^1\Pi_u$. This results in spin-orbit coupled potentials, and among them, the $(1)0_u^+$ and $(1)1_u$ potentials that asymptote to $^1S_0 + {}^3P_1$ are at the core of our experimental work so far. This *ab initio* model was later refined based on measurements of deeply bound $(1)1_u$ states in our experiment (see Sects. 3.6 and 4.2.1), and fitted to the Morse/Long-range form [Fig. 2.2b].

Deeply bound rovibrational states of $(1)1_u$ that lie below the potential minimum of $(1)0_u^+$ are mostly of $a^3\Sigma_u^+$ character. Mid-to-weakly bound states existing in the overlap region of $(1)0_u^+$ and $(1)1_u$ may be subject to **Coriolis coupling** which mixes different Ω states—this is especially so for near degenerate states such as the well-known pair that are bound by $\approx$8.3 GHz in $^{88}Sr_2$. As deeply bound states of $(1)1_u$ have few decay channels, they are predicted to have relatively narrow linewidths $\sim$ 10 kHz (although experimental observations have yet to confirm this).

$(1)0_u^+$ is the lower branch of the avoided crossing between $A^1\Sigma_u^+$ and $c^3\Pi_u$ [Fig. 2.2b]. Deeply bound rovibrational states of $(1)0_u^+$ inherit a large component from $A^1\Sigma_u^+$ (asymptoting to $^1S_0 + {}^1D_2$), and thus have singlet-singlet-like transitions to $X^1\Sigma_g^+$ states. On the other hand, weakly bound states of $(1)0_u^+$ have dominant $c^3\Pi_u$ character and possess narrower linewidths. The situation is complicated at the avoided crossing, but theory predicts alternating singlet and triplet characters for successive vibrational states [7]. For transitions between deeply bound $X^1\Sigma_g^+$ states and weakly bound $(1)0_u^+$ states, calculations reveal that $(2)0_u^+$ plays an important role as its depth extends below $^1S_0 + {}^3P_1$. A three-channel model consisting of $(1)0_u^+$, $(1)1_u$, and $(2)0_u^+$ reproduces the observed order of magnitude of these transition strengths [10].

Shallow-to-shallow transitions realize the simplest case of Dicke **subradiance** and **superradiance**. Very close to the dissociation threshold, we may approximate the molecular electronic wavefunction as the tensor product of two atomic electronic wavefunctions. For the ground state, this is simple: $|\mathrm{Sr}_2\rangle_g \approx |^1S_0\rangle|^1S_0\rangle$. For the excited state, there are two ways to symmetrize such a wavefunction: $|\mathrm{Sr}_2^*\rangle_{(g/u)} \approx \left(|^1S_0\rangle|^3P_1\rangle \mp |^3P_1\rangle|^1S_0\rangle\right)/\sqrt{2}$. As alluded by the subscripts, these wavefunctions possess gerade (g) and ungerade (u) inversion symmetries.[7] Multiplying out the transition dipole matrix element $\langle \mathrm{Sr}_2|_g\, d_0^{(1)}\, |\mathrm{Sr}_2^*\rangle_{(g/u)}$ we see that $\langle \mathrm{Sr}_2|_g\, d_0^{(1)}\, |\mathrm{Sr}_2^*\rangle_g \approx 0$ and $\langle \mathrm{Sr}_2|_g\, d_0^{(1)}\, |\mathrm{Sr}_2^*\rangle_u \approx \sqrt{2} \times \langle ^1S_0|d_0^{(1)}|^3P_1\rangle$. Therefore, $g \rightarrow g$ transitions from the ground state are quasi-forbidden, whereas $g \rightarrow u$ transitions from the ground state are enhanced by a factor of $\sqrt{2}$ (hence the transition rate will be twice as fast, $|\sqrt{2}|^2 = 2$). This is exactly Dicke subradiance and superradiance, respectively. For this reason, weakly bound excited states correlating to $^1S_0 + {}^3P_1$ with gerade symmetry $(0_g^+, 1_g)$ are subradiant, while those with ungerade symmetry $(0_u^+, 1_u)$ are superradiant (with roughly twice the atomic linewidth). Subradiant transitions to $(1)1_g$ from $X^1\Sigma_g^+$ have been observed in weakly bound dimers of both Yb_2 and Sr_2 [12], with narrow linewidths as small

[7] Inversion results in $|^1S_0\rangle|^3P_1\rangle \rightarrow (-1) \times |^3P_1\rangle|^1S_0\rangle$ as the P state has odd parity.

as <100 Hz limited by predissociation to $^3P_0$. Gerade $(1)0_g^+$ states may have even narrower linewidths, but these are yet to be found despite substantial experimental effort.

For the excited states, **J** is not solely due to nuclear rotations. From the argument in the previous subsection, we see that J **must be odd for** $(1)0_u^+$ of the bosonic dimer. In contrast, for $|\Omega| = 1$, the parity operator acts differently on the rotational wavefunctions, and so J **can take on any positive integer for** $(1)1_u$ (i.e., $J \neq 0$ or else it would imply $\Omega = 0$, which is not the case for 1_u).

Selection rules originate from the Wigner-Eckart theorem. The **electric dipole** $(E1)$ **selection rules** for an allowed molecular transition are $\Delta J = 0, \pm 1$ except $J = 0 \to J' = 0$, $|\Delta\Omega| \leq 1$, $\Delta M = 0, \pm 1$ depending on the probe laser polarization, and $g \leftrightarrow u$ for homonuclears. These selection rules may be broken in the presence of *mixed quantization*, where there is competition between the externally applied magnetic field and the electric field of the trapping laser in establishing a quantization axis [13]. Note that $J = 0 \to J' = 0$ is strictly forbidden for all multipoles of the driving field, and that $\Delta M \neq 0$ for $M = 0$ if $\Delta J = 0$. For more details on alkaline-earth metal molecular structure or angular momentum theory, see Refs. [14–17].

2.3 Photoassociation and Photofragmentation

2.3.1 Narrow-Line Photoassociation (State Preparation)

Photoassociation (PA) occurs when two particles scatter into a bound molecular state under the assistance of light. Laser light tuned near a PA resonance dramatically affects the interparticle scattering length, creating an optical Feshbach resonance. In the context of this thesis, one-photon PA binds two colliding $^1S_0$ strontium atoms into an excited rovibrational molecular state near the $^1S_0 + {}^3P_1$ threshold. It is succinctly described by the chemical equation:

$$\mathrm{Sr} + \mathrm{Sr} + \gamma \to \mathrm{Sr}_2^*,$$

where γ represents the photon and the (*) superscript implies that the molecule is electronically excited.

As explained in the previous subsection, weakly bound molecular states close to the $^1S_0 + {}^3P_1$ threshold (and with ungerade state symmetry for homonuclears) have natural linewidths that are approximately twice that of the atomic $^3P_1$, resulting in a spectrum of narrow and isolated lines below the intercombination, first observed in the seminal work of Zelevinsky et al. [18]. For this reason, laser-based PA via spontaneous emission in alkaline-earth metal gases can be performed with reduced off-resonant atomic scattering losses compared to alkali metals.

Successful PA is indicated by the reduction[8] of atoms in the trap as the PA laser is scanned across the PA resonance [19, 20]. The excited molecule subsequently decays, and the population is redistributed depending on the branching ratios given by Fermi's golden rule (transitions with larger transition strengths have larger branching ratios). If the dominant decay channel involves a bound molecular state in $X^1\Sigma_g^+$, then PA results in efficient ground state molecule formation. A simple kinematic argument indicates that the photoassociated molecules will have $\approx 2\times$ larger temperatures than the initial atomic cloud.

Bound-to-bound transition strengths are proportional to the square of the overlap integral of the vibrational wavefunctions called the Franck-Condon factor (FCF); see Sect. 4.2.1. The bonding between S and P states is usually of the resonant dipole type with a long-range potential varying as $\sim C_3/R^3$ plus additional terms. In contrast, two S states are held together by van der Waal forces with long-range behavior $\sim C_6/R^6$. The narrow linewidth (Γ_a) of $^3P_1$ facilitates ground state molecule production because $C_3 \propto \Gamma_a$ is small, so the long-range part of both excited and ground potentials scale predominantly as $\sim 1/R^6$, leading to good Franck-Condon overlap between weakly bound ground and excited states (i.e., large FCF). To list a few examples in $^{88}Sr_2$, PA to $(1)1_u(v' = -1, J' = 1)$ or $(1)0_u^+(-4, 1)$ efficiently creates the least bound vibrational states $X(v = -1, J = 0, 2)$, while PA to the adjacent $(1)1_u(-2, 1)$ or $(1)0_u^+(-5, 1)$ efficiently creates $X(v = -2, J = 0, 2)$. Negative values of the vibrational quantum numbers (v and v') imply the convention where we count downward, starting with -1, from the respective dissociation thresholds.

In this thesis, we perform PA after the atoms are loaded into a one-dimensional (1D) optical lattice from the red MOT. The PA laser is coaligned axially with the lattice, and the typical PA laser beam intensity and pulse duration are $\sim 15\,\mathrm{W/cm^2}$ and 2 ms. Care should be taken for the choice of PA parameters. Even though the PA laser is red detuned from the atomic resonance, off-resonant **scattering** eventually diminishes the overall molecule production efficiency, and laser-induced dipole forces may cause **breathing modes** in the gas. Moreover, long PA durations risk **contaminating** the sample with molecules in the unwanted (and undetected) vibrational states or stretched rotational sub-levels, as the atoms are continuously cycled through the excited state.

Even in an ideal situation, PA to $J' = 1$ creates a mixture of $J = 0, 2$ ground state molecules. Theoretically, for $\Omega' = 1 \rightarrow \Omega = 0$ the rotational factor (see Sect. 4.2.1) results in the decay from $J' = 1$ to $J = 0$ being two times more likely than to $J = 2$. The opposite is true for $\Omega' = 0 \rightarrow \Omega = 0$, which favors decay to $J = 2$ over $J = 0$. Therefore we expect the ratio of $J = 0$ to $J = 2$ ground state molecules to be distributed 2:1 if we PA using a 1_u state, and 1:2 using a 0_u^+ state. This has to be weighed against practical considerations, as 1_u states are

[8] The on-resonance loss curve should look non-exponential; i.e., $dn/dt = -k_2 n^2$.

generally more sensitive to changes in the magnetic field than 0_u^+. In either case, we can prepare a high purity $J = 0$ sample by photodissociating[9] the unwanted $J = 2$ molecules at (or above) the $^1S_0 + {}^3P_1$ threshold[10] with a weak laser frequency component added to the PA laser (via an acousto-optic modulator or electro-optic modulator).

2.3.2 *State-Selective Photofragmentation (Detection)*

Photofragmentation (also called photodissociation) is the process where a molecule is broken apart into its constituents by light. The angular distribution of the fragments contains plentiful information about the original molecular state and the continuum channel(s). The level of control afforded by ultracold molecule experiments has led to several new insights discovered by our group. A detailed overview of photodissociation (PD) and the novel features exhibited in the ultracold regime can be found in previous theses [10, 21, 22] and publications [20, 23–26].

The current work uses PD as a state-selective detection scheme for generating a spectroscopic signal. One-photon PD of ground state $X^1\Sigma_g^+$ molecules above the $^1S_0 + {}^3P_1$ dissociation threshold may be thought of as the time-reversed process of PA:

$$\mathrm{Sr}_2 + \gamma \rightarrow \mathrm{Sr} + \mathrm{Sr}^* + \epsilon.$$

The excess energy ϵ above the threshold is equally distributed and carried away as kinetic energy by the atoms. That is, $\epsilon = \frac{1}{2}m_{\mathrm{Sr}}v_{\mathrm{Sr}}^2 + \frac{1}{2}m_{\mathrm{Sr}}v_{\mathrm{Sr}}^2 = m_{\mathrm{Sr}}v_{\mathrm{Sr}}^2$, where m_{Sr} is the mass of a strontium atom and v_{Sr} is the speed of the atomic photofragment after dissociation. Note that ϵ is equivalent to the frequency difference of the dissociation laser with respect to the threshold multiplied by the Planck constant; i.e., it is set by the laser frequency. By performing near-threshold ($\epsilon \approx 0$) dissociation of a specific rovibrational ground state, the resulting slow-moving atomic fragments can be absorption imaged with high-fidelity [Fig. 2.3] via the $^1S_0 \leftrightarrow {}^1P_1$ cycling transition at 461 nm, providing a state-sensitive signal directly correlated with the number of molecules that were present in the trap. Molecules in other states are either not photodissociated because $\epsilon < 0$, or have fragments spatially separated from the spectroscopic signal as they emerge at much higher speeds.

---

[9] Conversely, clearing $J = 0$ without photodissociating $J = 2$ is not possible, unless through an excited state that has binding energy (wrt $^1S_0 + {}^3P_1$) greater than the rotational splitting of $J = 0, 2$.

[10] Again, recycling these molecules through a rovibrational excited state risks vibrational contamination or optical pumping into dark stretched rotational sub-levels.

For electric dipole ($E1$) PD of $J = 0$ ground states, it can be easily shown (either classically [6] or quantum mechanically [10]) that the photofragment angular distribution takes the simple form[11]

$$\sigma(\theta, \phi) = 1 + \beta_{20} P_2(\cos\theta), \tag{2.1}$$

where θ is the polar angle relative to the dissociation laser polarization (or whatever defines the quantization axis for the process), ϕ is the azimuthal angle, $P_2(x) \equiv (3x^2 - 1)/2$ is the second Legendre polynomial, and $-1 \le \beta_{20} \le 2$ is the anisotropy parameter that depends on ϵ and the transition strengths to the outgoing channels $(1)0_u^+$ and $(1)1_u$. We note that $\beta_{20} = +2$ for $\Delta\Omega = 0$ (parallel), while $\beta_{20} = -1$ for $|\Delta\Omega| = 1$ (perpendicular).

For parallel dissociation, the polarization of the dissociation laser is essentially imprinted onto the direction of the emerging photofragments (after a finite expansion time). Due to the presence of a repulsive[12] barrier of roughly 20 MHz for $(1)1_u$ arising from the resonant dipole interaction ($\sim C_3/R^3$), we expect near-threshold PD to be mostly parallel involving just the $X \to 0_u^+$ channel. However, if ϵ is above the barrier, PD changes to perpendicular as the $X \to 1_u$ channel is stronger for weakly bound X states.

We may model the optical density (OD) of the slowing moving photofragments [Fig. 2.3b] as a 2D Gaussian,

$$OD(a, b) = OD_{\max} \exp\left[\frac{-(a - a_0)^2}{2\sigma_a^2}\right] \exp\left[\frac{-(b - b_0)^2}{2\sigma_b^2}\right] + \text{background},$$

where a and b are arbitrarily chosen orthogonal axes in the fitting of the absorption images. To reduce statistical variation on $OD_{\max}$ and the extracted widths of the cloud (σ_a, σ_b), in the fits we fix the center points (a_0, b_0) using instances where the cloud is imaged with high optical density as a reference.

When we employ $OD_{\max}$ as a **spectroscopic signal** that is proportional to the total number of molecules in the trap, the dissociation time (t_{dissoc}), (a_0, b_0), *and* (σ_a, σ_b) are kept fixed. By taking the integral $\int\int \mathrm{d}a\mathrm{d}b\, OD(a, b)$, using the definition of optical density and the Beer-Lambert law, it is possible to express the number of recovered *atoms* (N_{at}) in terms of the fit parameters,

$$N_{\text{at}} = \frac{2\pi\sigma_a\sigma_b}{\sigma_{\text{res}}} OD_{\max}, \tag{2.2}$$

11 Assuming that the dissociation laser has linear polarization aligned to the quantization axis. Alternatively, at zero B-fields and large Rabi frequencies, the polarization of the dissociation laser defines the PD quantization axis.

12 The resonant dipole interaction is attractive and approximately twice as large in the case of $(1)0_u^+$ [18].

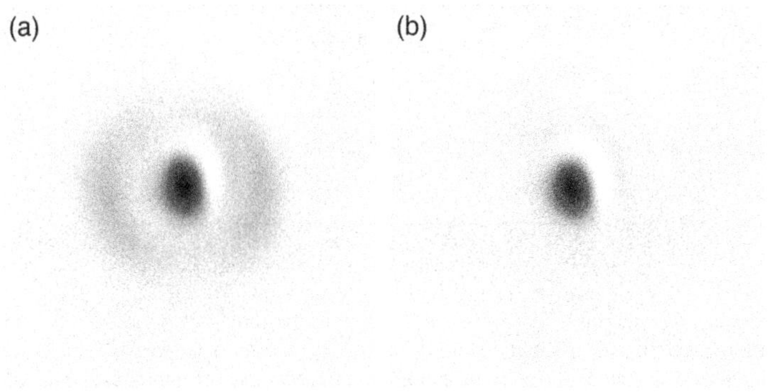

Fig. 2.3 Absorption images of slow-moving photofragments of lattice-confined $X(v = 62)$ molecules (with line-of-sight nearly along the 1D lattice axial direction) after a small expansion time ($\sim 150\,\mu s$) and almost-nulled magnetic field. The dissociation laser is tuned closely to $X(62, 0) \to {}^1S_0 + {}^3P_1$. (**a**) Possessing a smaller binding energy, $J = 2$ molecules emerge at greater speeds than $J = 0$ molecules, forming blurry outer rings with an angular distribution reminiscent of a dipole. (**b**) Wiping $J = 2$ molecules with an additional frequency component on the PA laser leaves a purified sample of $J = 0$ molecules. Suppose its atomic fragments (dark spot) are allowed to expand further either by waiting a long time before imaging or increasing ϵ. In that case, the image will reveal the $J = 0$ photofragment angular distribution, which also resembles a dipole (not shown, but see Ref. [24])

where $\sigma_{\text{res}} \equiv 3\lambda_{\text{res}}^2/(2\pi) \approx 1.014 \times 10^{-13}\,\text{m}^2$ is the resonant absorption cross-section for $\lambda_{\text{res}} = 461$ nm light. Note that to convert the widths (σ_a, σ_b) from units of image pixels to units of physical length, the magnification of the camera system needs to be measured (e.g., by observing a free-falling cloud of atoms and using the known local gravitational acceleration).

Importantly, N_{at} is *twice* the total number of diatomic molecules (N_{mol}) multiplied by the dissociation efficiency (η_{dissoc}). To estimate the dissociation efficiency for a given dissociation duration (t_{dissoc}) and laser intensity used for the photodissociation laser, we freely fit $OD_{\text{max}}, \sigma_a, \sigma_b$ as the dissociation pulse time is varied [Fig. 2.4]. As expected, the signal proportional to N_{at} has the form

$$y(t) = y(\infty)\left[1 - \exp(-R_{\text{dissoc}}\,t)\right],$$

where $y(\infty)$ and R_{dissoc} (the dissociation rate) are fitted parameters. Therefore,

$$\eta_{\text{dissoc}} = y(t_{\text{dissoc}})/y(\infty).$$

Putting it all together, the total number of *molecules* in the trap is

$$N_{\text{mol}} = \frac{N_{\text{at}}}{2\eta_{\text{dissoc}}} = \frac{\pi\sigma_a\sigma_b}{\sigma_{\text{res}}\eta_{\text{dissoc}}}\,OD_{\text{max}}. \tag{2.3}$$

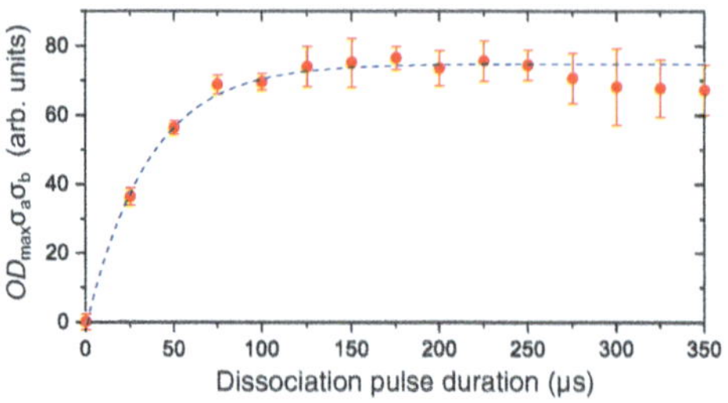

Fig. 2.4 Signal proportional to the recovered atom number (red points) versus photodissociation pulse duration at fixed laser intensity. Here, $(OD_{max}, \sigma_a, \sigma_b)$ are free parameters. Due to thermal effects, a small excess kinetic energy, ϵ, is inevitably imparted, and the cloud distribution eventually changes from Gaussian to a dipolar angular distribution as the fragments separate. To accurately determine the dissociation efficiency for the operational pulse duration, we fit (dashed blue) only to short pulse durations where the cloud resembles a singly peaked Gaussian. Typical dissociation pulse durations for $X(62, 0)$ state detection are $< 150\,\mu s$ at $\sim 10\,W/cm^2$ dissociation laser intensities

2.4 Optical Lattice Trap

2.4.1 *The Optical Dipole Force*

The optical dipole force arises from spatial gradients in laser intensity [27]. An external electromagnetic wave can polarize a neutral particle. For an infinitesimal change in the electric field at time t, $d\boldsymbol{\mathcal{E}}(t)$, the infinitesimal change in the induced electric dipole moment is $d\boldsymbol{d}_{ind}(t) = \alpha d\boldsymbol{\mathcal{E}}(t)$, where α is the polarizability (assumed here to be a scalar for simplicity) of the particle at the wavelength of the electromagnetic field. This induced dipole, in turn, interacts with the oscillating field that created it, leading to a time-varying potential for the particle,

$$U(t) = -\alpha \int_0^{\mathcal{E}} \boldsymbol{\mathcal{E}}'(t) \cdot d\boldsymbol{\mathcal{E}}'(t) = -\frac{1}{2}\alpha\mathcal{E}^2(t).$$

Typically, the field oscillates at near-optical frequencies, so we consider the cycle-averaged

$$U = -\frac{1}{2}\alpha \left\langle \mathcal{E}^2(t) \right\rangle_{\text{time average}}. \tag{2.4}$$

For example, if $\boldsymbol{\mathcal{E}}(t) = \hat{\mathbf{Z}}\mathcal{E}_0 \cos(\omega t)$, then $|U| = U_0 = \alpha\mathcal{E}_0^2/4$ where the additional factor of $1/2$ is due to the time-averaging.

2.4.2 One-Dimensional Optical Lattice

An optical lattice is the interference pattern created by a forward propagating field $\mathcal{E}_{\rm fwd}$ and the retroreflected field $\mathcal{E}_{\rm ret}$. Choosing the $\hat{\mathbf{Y}}$-direction to be the direction of propagation,[13] and $y = 0$ to be the location of the focus of the Gaussian beams, the instantaneous electric fields are

$$\mathcal{E}_{\rm fwd}(y, r, t) = \frac{w_0}{w(y)} \exp\left(\frac{-r^2}{w^2(y)}\right) [\mathbf{a}\cos(ky - \omega t) - \mathbf{b}\sin(ky - \omega t)], \tag{2.5}$$

$$\mathcal{E}_{\rm ret}(y, r, t) = \frac{w_0}{w(y)} \exp\left(\frac{-r^2}{w^2(y)}\right) [\mathbf{a}\cos(ky + \omega t) + \mathbf{b}\sin(ky + \omega t)], \tag{2.6}$$

where we have ignored the Gouy phase and arbitrarily chosen the relative phase of the fields for calculation simplicity without loss of generality.[14] $k = 2\pi/\lambda_{\rm latt}$ and $\omega = 2\pi f_{\rm latt}$ are, respectively, the angular wavenumber and angular frequency of the lattice laser light. The radial distance $r \equiv \sqrt{x^2 + z^2}$. To be general, the polarization of the electric field is assumed to be elliptical, with semi-major axis $\mathbf{a}$ and semi-minor axis $\mathbf{b}$ (i.e., $\mathbf{a} \cdot \mathbf{b} = 0$). For example, if the field were linearly polarized along $\hat{\mathbf{Z}}$, then we may write $\mathbf{a} = \hat{\mathbf{Z}}\mathcal{E}_0$ and $\mathbf{b} = 0$ where $\mathcal{E}_0$ is the amplitude of the field. The $1/e^2$ Gaussian beam waist[15] evolves along y as

$$w(y) = w_0 \sqrt{1 + \left(\frac{2y}{kw_0^2}\right)^2}. \tag{2.7}$$

The superposition of the counter-propagating fields results in the standing wave

$$\mathcal{E}^2_{\rm tot}(y, r, t) = |\mathcal{E}_{\rm fwd}(y, r, t) + \mathcal{E}_{\rm ret}(y, r, t)|^2 \tag{2.8}$$

$$= 4\cos^2(ky) \frac{w_0^2}{w^2(y)} \exp\left(\frac{-2r^2}{w^2(y)}\right) \left[|\mathbf{a}|^2 \cos^2(\omega t) + |\mathbf{b}|^2 \sin^2(\omega t)\right]. \tag{2.9}$$

[13] So that the corresponding Pauli matrices retain their conventional form; i.e., $\sigma_x = \begin{pmatrix} 0 & 1 \\ 1 & 0 \end{pmatrix}$, $\sigma_y = \begin{pmatrix} 0 & -i \\ i & 0 \end{pmatrix}$, $\sigma_z = \begin{pmatrix} 1 & 0 \\ 0 & -1 \end{pmatrix}$. These are closely related to Jones calculus, a powerful framework for describing polarized light.

[14] In practice, the positions of the nodes and anti-nodes are determined by their distance relative to the retroreflecting mirror, which sets a boundary condition. Also, here, we are assuming no return power losses.

[15] Note the factor of 2 difference between the definition for w with the conventional (mathematical) Gaussian σ.

Taking the time average over one cycle ($2\pi/\omega$),

$$\left\langle \mathcal{E}_{\text{tot}}^2(y,r,t) \right\rangle_{\text{time averaged}} = 4\cos^2(ky)\frac{w_0^2}{w^2(y)}\exp\left(\frac{-2r^2}{w^2(y)}\right) \times \frac{1}{2}\left[|\mathbf{a}|^2 + |\mathbf{b}|^2\right]. \tag{2.10}$$

Using Eq. (2.4), we arrive at the form of a 1D optical lattice potential[16]

$$U(y,r) = -U_0\cos^2(ky)\frac{w_0^2}{w^2(y)}\exp\left(\frac{-2r^2}{w^2(y)}\right). \tag{2.11}$$

To relate the trap depth U_0 with the physical beam parameters, we first observe that the irradiance[17] of the forward pass beam at the focus is the time average of the Poynting flux,

$$\begin{aligned} I_{\text{fwd}}(r) &= \frac{1}{2}c\left\langle \epsilon_0\,|\boldsymbol{\mathcal{E}}_{\text{fwd}}(0,r,t)|^2 + \frac{1}{\mu_0}\,|\boldsymbol{\mathcal{B}}_{\text{fwd}}(0,r,t)|^2 \right\rangle_{\text{time average}} \\ &= \frac{1}{2}\epsilon_0 c \times 2\left\langle |\mathbf{a}|^2\cos^2(\omega t) + |\mathbf{b}|^2\sin^2(\omega t)\right\rangle_{\text{time average}} \times \exp\left(\frac{-2r^2}{w_0^2}\right) \\ &= \frac{1}{2}\epsilon_0 c \times \left[|\mathbf{a}|^2 + |\mathbf{b}|^2\right] \times \exp\left(\frac{-2r^2}{w_0^2}\right) \qquad (2.12) \\ &\equiv I_{\text{fwd},0}\exp\left(\frac{-2r^2}{w_0^2}\right), \qquad (2.13) \end{aligned}$$

where ϵ_0 (μ_0) is the vacuum permittivity (permeability), c is the speed of light in vacuum, and in the second step we used the fact that $\boldsymbol{\mathcal{E}} = c\boldsymbol{\mathcal{B}}$ for a plane wave.[18] The total radiation power is the integral of the irradiance over r,

$$P_{\text{latt}} = \int_0^\infty dr\, 2\pi r I_{\text{fwd}}(r) = I_{\text{fwd},0}\frac{\pi w_0^2}{2}. \tag{2.14}$$

We note that P_{latt} corresponds to the power measured by a power meter in the path of the forward pass beam. Comparing Eqs. (2.12) and (2.14), we find

---

[16] Ignoring gravity. In lattice clocks, the lattice is usually oriented along gravity to suppress tunneling. It is horizontal for practical reasons for the experiments in this thesis (and prior).

[17] Colloquially, irradiance is also called the "intensity". See Ref. [28] for a careful discussion of numerical factors.

[18] This is a good approximation near $y = 0$.

$$I_{\text{fwd},0} = \frac{2P_{\text{latt}}}{\pi w_0^2} = \frac{1}{2}\epsilon_0 c \times \left[|\mathbf{a}|^2 + |\mathbf{b}|^2\right], \tag{2.15}$$

which for linear (or pure circular) polarization has the familiar form $I_{\text{fwd},0} = \epsilon_0 c \mathcal{E}_0^2/2$.

Substituting Eqs. (2.15) into (2.10), we therefore identify the lattice depth to be

$$U_0 = \frac{4\alpha P_{\text{latt}}}{\pi w_0^2 \epsilon_0 c}. \tag{2.16}$$

We note that the local intensity at the anti-nodes is $4 \times I_{\text{fwd},0}$ due to constructive interference.

2.4.3 Tight Confinement in a 1D Lattice

Let us theoretically examine the quantum mechanics of a particle of mass M in an optical lattice. Near the focus ($y \approx 0$), we may approximate

$$U(y,r) \approx -U_0 \cos^2(ky) \exp\left(\frac{-2r^2}{w_0^2}\right) \equiv -U_0(r)\cos^2(ky). \tag{2.17}$$

Concentrating on the axial (y) direction, the energy levels (E_n) are obtained by solving the time-independent Schrödinger equation

$$\left[-\frac{\hbar^2}{2M}\frac{\partial^2}{\partial y^2} + U(y,r)\right]\phi_n(y) = E_n \phi_n(y). \tag{2.18}$$

Since $U(y,r)$ is periodic along y with lattice sites ("pancakes") separated every $\pi/k = \lambda_{\text{latt}}/2$, Bloch's theorem tells us that the wavefunctions take the form

$$\phi_n(y) = e^{iqy} u_{n,q}(y), \tag{2.19}$$

where $\hbar q$ is the crystal momentum and $u_{n,q}(y)$ are functions with periodicity[19] $\lambda_{\text{latt}}/2$ labeled by the band index n. For this potential, $u_{n,q}(y)$ can be calculated rather straightforwardly. To do so, we notice that the effect of the momentum operator ($\mathbf{p}_{\text{op},y}$) on $\phi_n(y)$ is

$$\mathbf{p}_{\text{op},y}\phi_n(y) = -i\hbar\frac{\partial}{\partial y}\left[e^{iqy}u_{n,q}(y)\right] = e^{iqy}\left[\hbar q + \mathbf{p}_{\text{op},y}\right]u_{n,q}(y). \tag{2.20}$$

[19] That is, $u_{n,q}(y) = u_{n,q}(y + \lambda_{\text{latt}}/2)$

Therefore, the Schrödinger equation (2.18) reduces to

$$\left[\frac{1}{2M}\left(\hbar q + \mathbf{p}_{\mathrm{op},y}\right)^2 + U(y,r)\right] u_{n,q}(y) = E_{n,q} u_{n,q}(y), \tag{2.21}$$

where we add the subscript q to remind ourselves that the energies depend on q (i.e., energy bands will form in momentum space). Next, the periodicity of $u_{n,q}(y)$ and $U(y,r)$ allows us to expand them as Fourier series (up to a normalization constant),

$$u_{n,q}(y) = \sum_l c_l e^{i(2lky)}, \tag{2.22}$$

$$U(y,r) = \sum_m V_m e^{i(2mky)}, \tag{2.23}$$

with l, m being integers. Our goal now is to find the Fourier coefficients c_l.

For our potential

$$U(y,r) = -U_0(r)\cos^2(ky) = -U_0(r)\left[e^{i(2ky)} + e^{-i(2ky)} + 2\right]/4. \tag{2.24}$$

Hence, in Eq. (2.23) only $V_{+1} = V_{-1} = -U_0(r)/4$ and $V_0 = -U_0(r)/2$ are non-zero. Intuitively, this makes sense: a lattice with periodicity $2\pi/(2k)$ will scatter momentum $\pm 2k$. Substituting both Fourier series into Eq. (2.21) and comparing Fourier terms, we find

$$\left[E_{\mathrm{r}}(q/k + 2l)^2 + V_0\right] c_l + V_{+1} c_{l-1} + V_{-1} c_{l+1} = E_{n,q} c_l, \tag{2.25}$$

where $E_{\mathrm{r}} \equiv \hbar^2 k^2/(2M) = h^2/(2M\lambda_{\mathrm{latt}}^2)$ is called the recoil energy (it serves as a natural unit to express trap energies).

Equation (2.25) can be recast as a matrix equation,

$$\tilde{\boldsymbol{h}}\,\mathbf{c} = E_{n,q}\mathbf{c}, \tag{2.26}$$

where $\tilde{\boldsymbol{h}}$ is an infinite-dimensional, tri-diagonal square matrix that has the form

$$\tilde{h} = \begin{pmatrix} \ddots & V_{-1} & 0 & 0 & 0 \\ V_{+1} & E_{\mathrm{r}}(q/k + 2(l-1))^2 + V_0 & V_{-1} & 0 & 0 \\ 0 & V_{+1} & E_{\mathrm{r}}(q/k + 2l)^2 + V_0 & V_{-1} & 0 \\ 0 & 0 & V_{+1} & E_{\mathrm{r}}(q/k + 2(l+1))^2 + V_0 & V_{-1} \\ 0 & 0 & 0 & V_{+1} & \ddots \end{pmatrix}.$$

Similarly, the vector of Fourier coefficients $\mathbf{c}$ is

$$\mathbf{c} = \begin{pmatrix} \vdots \\ c_{l-1} \\ c_l \\ c_{l+1} \\ \vdots \end{pmatrix}.$$

Algebraically, we may also write

$$\tilde{h}_{l,l'} = \begin{cases} E_r(q/k + 2l)^2 + V_0, & \text{if } l = l' \\ V_{\pm 1}, & \text{if } |l - l'| = 1 \\ 0, & \text{otherwise} \end{cases} \tag{2.27}$$

Diagonalizing $\tilde{\boldsymbol{h}}$ returns the eigenvalues $E_{n,q}$ and the corresponding eigenvectors $\mathbf{c}$. For each $\mathbf{c}$, the Bloch wavefunction is constructed using Eqs. (2.22) and (2.19). In practice, most coefficients in $\mathbf{c}$ are negligible. A deeper trap necessitates a larger number of Fourier components for an accurate calculation. For $U_0 \sim 500\,E_r$, we truncate $\tilde{\boldsymbol{h}}$ to about 40×40. The energy width of the n-th band is the difference of the maximum and minimum value of $E_{n,q}$ as a function of q.

Figure 2.5 shows a plot of $|u_{1,0}|^2$ for various trap depths U_0. As U_0 increases, the probability density localizes, and the local wavefunction (called the Wannier function, the Fourier inverse of the Bloch wavefunctions) at each site approaches that of the harmonic oscillator. Thus a deep optical lattice can be approximated as an array of harmonic microtraps. Since the positions of the lattice anti-nodes are set by the position of the retro-reflecting mirror, which is clamped onto a massive body, spectroscopy in deep lattices will be recoil-free (analogous to the Mössbauer effect). For weak confinement, the width of the energy bands leads to decoherence and peak broadening effects in clock spectroscopy, and tunneling can open up s-wave collisions even with fermions by making them distinguishable. Further information can be found in literature (e.g., Refs. [29, 30]).

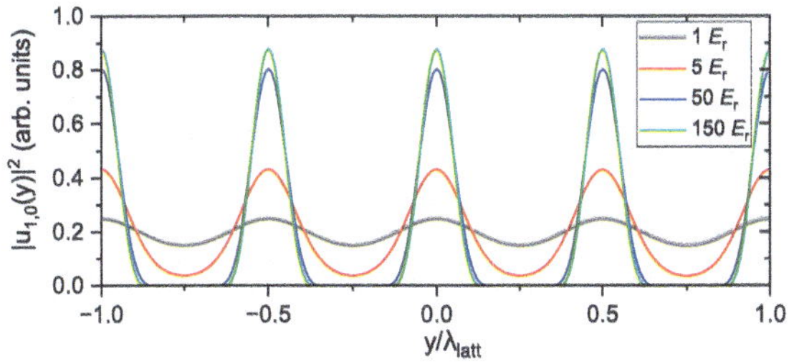

Fig. 2.5 Ground state probability density of a particle in an idealized optical lattice shown for 5 lattice sites. Here, $n = 1, q = 0$. Localization occurs as the trap depth increases, and the local wavefunction at each lattice site approaches that of a simple harmonic oscillator

2.4.4 Carrier and Sideband Transitions, the Lamb-Dicke Regime

Having established that an array of harmonic microtraps is a good approximation for the sites of a deep optical lattice, we expand Eq. (2.17) to leading order for the central microtrap (i.e., around $y \approx 0$ and $r \approx 0$),

$$U(y,r) = -U_0 \cos^2(ky) \exp\left(\frac{-2r^2}{w_0^2}\right)$$

$$\approx U_0 \left(-1 + k^2y^2 + \frac{2r^2}{w_0^2} - \frac{2r^2}{w_0^2}k^2y^2 - \frac{k^4y^4}{3} - \frac{2r^4}{w_0^4} + \dots\right), \tag{2.28}$$

$$\approx U_0 k^2 y^2 + \frac{2U_0}{w_0^2} r^2 - U_0 + \dots. \tag{2.29}$$

Making the correspondence to the harmonic potential and ignoring the cross terms, we thus define the axial and radial trap angular frequencies

$$2\pi f_{\mathrm{ax}} \equiv \omega_{\mathrm{ax}} \equiv \sqrt{\frac{1}{M}\frac{\partial^2 U(y,r)}{\partial y^2}} = k\sqrt{\frac{2U_0}{M}}, \tag{2.30}$$

$$2\pi f_{\mathrm{rad}} \equiv \omega_{\mathrm{rad}} \equiv \sqrt{\frac{1}{M}\frac{\partial^2 U(y,r)}{\partial r^2}} = \frac{2}{w_0}\sqrt{\frac{U_0}{M}}. \tag{2.31}$$

A related quantity is $f_{\mathrm{ax}}/f_{\mathrm{rad}} = \pi\sqrt{2}(w_0/\lambda_{\mathrm{latt}})$ which is the aspect ratio of a microtrap.[20] The three quantities f_{ax}, f_{rad} and λ_{latt} completely specifies the trap geometry. The trap depth may be expressed in terms of the axial trap frequency as

$$U_0 = \frac{1}{2} M f_{\mathrm{ax}}^2 \lambda_{\mathrm{latt}}^2, \tag{2.32}$$

$$U_0/E_{\mathrm{r}} = M^2 f_{\mathrm{ax}}^2 \lambda_{\mathrm{latt}}^4 / h^2. \tag{2.33}$$

If the trap frequencies are obtained using a sample of molecules, then $M = 2m_{\mathrm{Sr}}$ is the mass of the dimer, and Eq. (2.32) gives the trap depth for the molecules.

Figure 2.6 shows the energy diagram for a particle in a harmonic trap. Interrogation of the particle's internal states may alter its motional quantum number (n_{ax} for the axial direction, n_{rad} for the radial direction[21]). A transition that leaves n

[20] Since $w_0 \gg \lambda_{\mathrm{latt}}$, the sites of a 1D lattice are often described as "pancakes".

[21] Note for each value of n_{rad}, there are $n_{\mathrm{rad}}+1$ degenerate radial motional states; c.f., 2D harmonic oscillator.

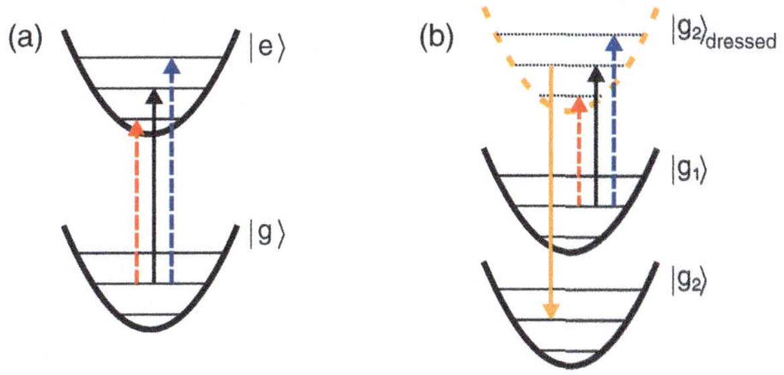

Fig. 2.6 Each black curve represents the external potential that traps a molecule in a given rovibronic state. Tightly confined molecules have quantized motional trap states (horizontal lines, shown for $n = 0, 1, 2$ in increasing energy) that can be separately addressed in the resolved sideband regime. (**a**) The molecule makes a rovibronic transition from the initial molecular state (bottom curve) to the final molecular state (top curve), mediated by the absorption of a single photon. The arrows represent the type of motional state transition, shown here for a molecule starting in $n = 1$. Red arrow, red sideband ($\Delta n = -1$); black arrow, carrier ($\Delta n = 0$); blue arrow, blue sideband ($\Delta n = +1$). (**b**) For Raman transitions, one way of interpreting sideband spectroscopy is to think of the final molecular state that is coupled by the downleg (or anti-Stokes) as being dressed (upper dotted curve and lines). Carrier and sideband transitions occur, as in (**a**), when the upleg is scanned across the Raman resonance

unchanged is called a carrier transition, while a transition that changes n by $+1$ (-1) is called the first blue (red) sideband transition. Suppose the probed transition is much narrower than the separation of the sidebands relative to the carrier. In that case, the spectroscopy is in the **resolved sideband regime**, permitting the direct measurement[22] of f_{ax} and f_{rad}.

In our experiment, we have so far observed two types of molecular transitions narrow enough for sideband spectroscopy; the $X \to 1_g$ subradiant transitions, and the $X \to X$ Raman transitions. In both cases, the transition rate is proportional to the square of the (effective) Rabi angular frequency (Ω_0) and a matrix element involving the motional states ($|n_{\text{ax}}, n_{\text{rad}}\rangle$),

$$\Omega^2 = \Omega_0^2 \left| \langle n'_{\text{ax}}, n'_{\text{rad}} | e^{i \Delta \mathbf{k}_{\text{prb}} \cdot \boldsymbol{\mathcal{R}}_{\text{op}}} | n_{\text{ax}}, n_{\text{rad}} \rangle \right|^2, \tag{2.34}$$

where $\hbar \Delta \mathbf{k}_{\text{prb}} = \hbar \{\Delta k_{\text{prb},x}, \Delta k_{\text{prb},y}, \Delta k_{\text{prb},z}\}$ is the net momentum imparted by the probe(s) in each direction and $\boldsymbol{\mathcal{R}}_{\text{op}} = \{\mathbf{x}_{\text{op}}, \mathbf{y}_{\text{op}}, \mathbf{z}_{\text{op}}\}$ is the position operator vector. We note that for two-photon transitions $|\Delta \mathbf{k}_{\text{prb}}|$ is dependent on the relative angle between the probes, and whether it is a Λ (Raman) or a Ξ (ladder) system; see also

[22] An alternate method involves observing the parametric heating resonances of the molecules by modulating the trap depth (e.g., modulating the lattice laser power using an AOM). In principle, trap losses will happen at twice the sideband frequencies, but sometimes subharmonics or higher harmonics may occur. This method does not require narrow transitions or the resolved sideband regime.

Eqs. (3.36) and (4.2). Following Ref. [31], the matrix element can be expanded in terms of the generalized Laguerre polynomials, $L_n^{(\alpha)}(\eta^2)$,

$$\langle n'_{\mathrm{ax}}, n'_{\mathrm{rad}}|e^{i\Delta\mathbf{k}_{\mathrm{prb}}\cdot\boldsymbol{\mathcal{R}}_{\mathrm{op}}}|n_{\mathrm{ax}}, n_{\mathrm{rad}}\rangle = \prod_{j\in\{x,y,z\}} \langle n'_j|e^{i\eta_j(\hat{a}_j+\hat{a}_j^\dagger)}|n_j\rangle, \tag{2.35}$$

$$= \prod_{j\in\{x,y,z\}} e^{-\eta_j^2/2}\eta_j^{|\Delta n_j|}\sqrt{\frac{(n_j^<)!}{(n_j^>)!}}\,L_{n_j^<}^{(|\Delta n_j|)}(\eta_j^2). \tag{2.36}$$

In the first step, the position operators for each direction ($j \in \{x, y, z\}$) are expressed as harmonic oscillator creation and annihilation operators, and we define the **Lamb-Dicke parameters**

$$\eta_j \equiv \Delta k_{\mathrm{prb},j}\sqrt{\frac{\hbar}{2M\omega_j}} = \frac{\Delta k_{\mathrm{prb},j}}{2\pi}\sqrt{\frac{h}{2Mf_j}}, \tag{2.37}$$

where the trap frequencies f_j are $f_x = f_z = f_{\mathrm{rad}}$ and $f_y = f_{\mathrm{ax}}$. In the second step, $n_j^>$ ($n_j^<$) is the greater (smaller) of the values n'_j and n_j, and $\Delta n_j \equiv n_j^> - n_j^<$. Spectroscopy along the axial direction is typically in the **Lamb-Dicke regime** (i.e., $\eta_y \equiv \eta_{\mathrm{ax}} < 1$) as the axial confinement is strong, but this is not the case along the weaker radial direction ($\eta_x = \eta_z \equiv \eta_{\mathrm{rad}} \gg 1$).

By writing out the Laguerre polynomials in Eq. (2.36) and keeping only leading order terms, we will ultimately find that sideband transitions are suppressed by a factor $\sim \eta_j^2$ relative to the carrier, and that the blue sideband will always be stronger than the red sideband (intuitively because $\hat{a}_j^\dagger|n_j\rangle = \sqrt{n_j+1}|n_j\rangle$ and $\hat{a}_j|n_j\rangle = \sqrt{n_j}|n_j\rangle$). For spectroscopy along the axial direction,

$$\Omega^2 \approx \begin{cases} \Omega_0^2, & \Delta n_{\mathrm{ax}} = 0 \text{ (carrier)}, \\ \Omega_0^2\,\eta_{\mathrm{ax}}^2\,(n_{\mathrm{ax}}+1), & \Delta n_{\mathrm{ax}} = +1 \text{ (blue sideband)}, \\ \Omega_0^2\,\eta_{\mathrm{ax}}^2\,n_{\mathrm{ax}}, & \Delta n_{\mathrm{ax}} = -1 \text{ (red sideband)}. \end{cases} \tag{2.38}$$

Figure 2.7 shows exemplary axial and radial sideband spectra excited using Raman clock transitions. To enhance the axial sideband spectra, the Raman probes counter-propagate along the axial direction to increase the Lamb-Dicke parameter.[23] To impart momentum along the radial direction and acquire the radial sidebands, one of the co-propagating Raman probes is intentionally misaligned from the other so that they cross at a small angle at the molecules.

[23] Counter-propagating Raman: $|\Delta k_{p,y}| = 2\pi/\lambda_\downarrow + 2\pi/\lambda_\uparrow$. Co-propagating Raman: $|\Delta k_{p,y}| = 2\pi/\lambda_\downarrow - 2\pi/\lambda_\uparrow$.

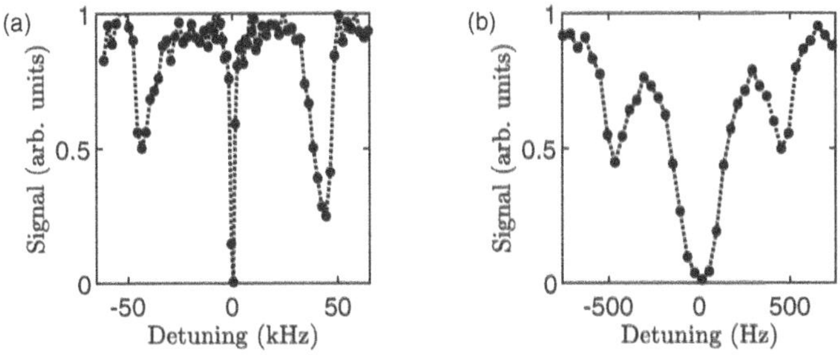

Fig. 2.7 (**a**) Axial sidebands excited by counter-propagating laser beams, interrogating the $X(62, 0) \to X(61, 0)$ Raman transition. This pair of states is naturally near-magic, permitting the trap depth of $X(62, 0)$ to be determined over a wide range of lattice wavelengths. (**b**) Radial sidebands excited by co-propagating laser beams, intentionally misaligned by a slight relative angle, driving the Raman transition $X(62, 0) \to X(0, 0)$ in a magic wavelength lattice. The upleg was stepped for both traces. The left, middle, and right peaks are the corresponding red sideband, carrier transition, and blue sideband. Black points are single shots of the experiment. The dotted lines are guides to the eye

The axial sidebands are asymmetric, with the sharp edges facing away from the carrier. This feature is due to the coupling between the weak radial confinement and the tight axial motion. Including the lowest order, axial-radial cross term $-r^2y^2(2U_0k^2/w_0^2)$ in Eq. (2.30) leads to a distortion (or shift) in the axial trap frequency as a function of r,

$$f'_{\mathrm{ax}}(r) \simeq f_{\mathrm{ax}}\sqrt{1 - \frac{2r^2}{w_0^2}} \quad \Rightarrow f'_{\mathrm{ax}}(r) - f_{\mathrm{ax}} \approx -\frac{f_{\mathrm{ax}}r^2}{w_0^2}.$$

Thus, the thermal probability distribution in r (and hence n_{rad}) smears out the observed axial sidebands resulting in a Boltzmann tail falling toward the carrier. Operationally, the sharp edges of the axial sidebands correspond very well to the expression in Eq. (2.30). A rigorous quantum mechanical derivation [32] obtains the following axial sideband lineshape (sharp edge centered at $\delta f = 0$, with the Boltzmann tail falling toward positive[24] frequency shifts $\delta f > 0$):

$$p_{\mathrm{sideband}}(\delta f) \propto \begin{cases} |\delta f|^3 \exp\left[-4U_0\,\delta f/(k_B T f_{\mathrm{ax}})\right], & \delta f > 0 \\ 0, & \text{otherwise} \end{cases} \tag{2.39}$$

where k_B is the Boltzmann constant, and T is the molecular ensemble temperature. Alternatively, it is possible to perform numerical fitting to the sidebands using the method in Refs. [33, 34].

[24] For a lineshape with a tail falling in the opposite direction, substitute $\delta f \to -\delta f$.

The ratio of the areas under the blue and red axial sidebands (A_{bsb}/A_{rsb}) can be utilized for thermometry.[25] For sideband thermometry, it suffices to fit skewed Gaussians to estimate the area under the curves. From Eq. (2.38), we see that

$$\frac{A_{bsb}}{A_{rsb}} = \frac{\langle n_{ax} \rangle + 1}{\langle n_{ax} \rangle}. \tag{2.40}$$

Straight-forward calculation using the partition function of a harmonic oscillator gives

$$\langle n_{ax} \rangle = \frac{1}{\exp\left(\frac{hf_{ax}}{k_B T}\right) - 1}. \tag{2.41}$$

Combining the formulas, we get

$$T = \frac{hf_{ax}}{k_B \ln\left(A_{bsb}/A_{rsb}\right)}. \tag{2.42}$$

Lastly, we note that the observed frequency separation ($2f_{sb}$) of the red and blue sidebands is not necessarily the same as $2f_{ax}$ when the probed molecular states have different polarizabilities (i.e., in a non-magic wavelength trap). To see this, consider the trap energies of a particle in its initial (unprimed) and final (primed) states:

$$E(n_{ax}, n_{rad}) \approx -U_0 + hf_{ax}\left(n_{ax} + \frac{1}{2}\right) + hf_{rad}\left(n_{rad} + 1\right), \tag{2.43}$$

$$E'(n'_{ax}, n'_{rad}) \approx -U'_0 + hf_{ax}\sqrt{\frac{\alpha'}{\alpha}}\left(n'_{ax} + \frac{1}{2}\right) + hf_{rad}\sqrt{\frac{\alpha'}{\alpha}}\left(n'_{rad} + 1\right), \tag{2.44}$$

where α, α' are the polarizabilities of the particle in the initial and final internal states respectively. Taking the axial sidebands as an example, half of the separation between the red ($n'_{ax} - n_{ax} = -1$) and blue ($n'_{ax} - n_{ax} = +1$) axial sidebands is

$$f_{sb} \approx f_{ax}\sqrt{\frac{\alpha'}{\alpha}}. \tag{2.45}$$

In practice, we measure f_{sb} for various P_{latt} to obtain the slope f_{sb}^2/P_{latt}. At the same time, by measuring the carrier line center at various P_{latt} (which shifts due to the polarizability difference), we get the *slope* of the differential light shift (L_0); see Fig. 2.8a and b. From Eq. (2.32),

[25] Sideband thermometry is more susceptible to molecule number and laser power fluctuations than carrier thermometry. Moreover, it only measures the axial thermal distribution $\langle n_{ax} \rangle$, whereas carrier thermometry samples both $\langle n_{ax} \rangle$, $\langle n_{rad} \rangle$.

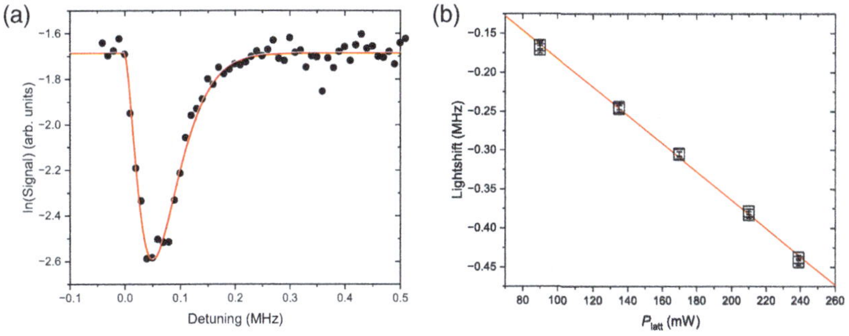

Fig. 2.8 (**a**) Vibrational clock transition $X(62, 0) \rightarrow X(0, 0)$ in a non-magic wavelength lattice far from any $X \rightarrow 1_u$ resonances ($\lambda_{\text{latt}} = 1005.6$ nm for this dataset). The Raman probe beams are co-propagating along the lattice and are detuning by one cavity FSR (approximately 1.5 GHz) from $(1)0_u^+(11, 1)$. This is closer to resonance than normal clock operating conditions to gain a larger effective Rabi frequency to drive the thermally broadened transition. The thermal tail of the carrier line shape can be utilized for accurate thermometry. Here, $\alpha'/\alpha > 1$ and the upleg laser frequency is stepped across the two-photon resonance, while that of the downleg is kept fixed. The spectroscopic signal is proportional to the number of $X(62, 0)$ molecules. Since the signal is depleted exponentially over the interrogation time, and the depletion rate more accurately reflects p_{carrier}, we take the *natural logarithm* of the signal before making the lineshape fit. Each black point is averaged over three consecutive experimental scans of the line. (**b**) Measured differential lattice light shift (black squares) due to the polarizability difference of $X(0, 0)$ and $X(62, 0)$. The carrier line positions (roughly the location of the sharp edge) at various lattice powers are determined from the carrier lineshape fit. Error bars denote 1σ statistical uncertainties. The slope of the linear fit is L_0. See main text for details

$$L_0 \equiv \frac{(U_0 - U_0')/h}{P_{\text{latt}}} = \frac{M\lambda_{\text{latt}}^2}{2h} f_{\text{ax}}^2 \left(1 - \frac{\alpha'}{\alpha}\right) \frac{1}{P_{\text{latt}}}. \tag{2.46}$$

Rearranging Eq. (2.46), we get

$$\frac{\alpha'}{\alpha} = 1 - \frac{2h}{M\lambda_{\text{latt}}^2} \frac{L_0}{f_{\text{ax}}^2/P_{\text{latt}}} = 1 - \frac{4E_r}{h} \frac{L_0}{f_{\text{ax}}^2/P_{\text{latt}}}. \tag{2.47}$$

Since the differential light shift $L_0 P_{\text{latt}}$ and f_{ax}^2 both depend linearly on lattice intensity, this method of expressing the polarizability ratio is agnostic to any geometric parameters that are difficult to access experimentally (e.g., lattice beam waist). Substituting Eq. (2.47) into Eq. (2.45), we find that the raw sideband frequency will have to be corrected by an additional term to give the trapping frequency of the initial state (a similar argument holds for f_{rad})

$$\frac{f_{\text{ax}}^2}{P_{\text{latt}}} \approx \frac{f_{\text{sb}}^2}{P_{\text{latt}}} + \frac{2hL_0}{M\lambda_{\text{latt}}^2}. \tag{2.48}$$

Because $f_{sb} \to f_{ax}$ as $L_0 \to 0$, we can circumvent the need to apply Eq. (2.48) by scanning a naturally near-magic Raman transition like $X(62, 0) \to X(61, 0)$. Note that for Raman transitions, the sign of L_0 is positive (negative) if the Raman line centers shift positively (negatively) for increasing values P_{latt} when the *upleg* laser frequency is scanned. The signs are reversed if the *downleg* is scanned instead.

2.4.5 Carrier Thermometry

A narrow carrier line is thermally broadened in a non-magic wavelength trap. Consider the value of $E'(n_{ax}, n_{rad}) - E(n_{ax}, n_{rad})$. In addition to the overall shift in central value [Eq. (2.46)], the frequency shifts $n_{ax} f_{ax}(\sqrt{\alpha'/\alpha} - 1)$ and $n_{rad} f_{rad}(\sqrt{\alpha'/\alpha} - 1)$ become compounded for increasingly higher motional states n_{ax}, n_{rad}.

The thermal probability distribution for n_{ax}, n_{rad} is therefore imprinted onto the carrier spectrum, resulting in an asymmetric lineshape that can be leveraged for highly accurate thermometry [32, 35]. A quantum mechanical derivation for the carrier lineshape [32] implies that (shape edge centered at $\delta f = 0$, and tail falling toward positive δf):

$$p_{carrier}(\delta f) \propto \begin{cases} (\delta f)^2 \exp(-B\delta f), & \delta f > 0 \\ 0, & \text{otherwise} \end{cases} \tag{2.49}$$

where

$$B \equiv \frac{h}{k_B T \left(\sqrt{\alpha'/\alpha} - 1\right)}. \tag{2.50}$$

For accurate carrier thermometry, the differential polarizability should be maximized (i.e., α'/α is as large as possible) while keeping other sources of broadening minimal (e.g., effects due to probe power, magnetic field, etc.).

As an illustrative example, Fig. 2.8a shows a carrier spectrum for the $X(62, 0) \to X(0, 0)$ Raman transition at a non-magic wavelength, fitted with Eq. (2.49). These states span the entire ground potential and have maximal polarizability difference, $\alpha' - \alpha$. The carrier fit also returns the position of the sharp edge, giving the differential light shift when plotted against various P_{latt}, as shown in Fig. 2.8b. The slope of the linear fit is L_0; see Eq. (2.46).

To invert Eq. (2.50) and extract the ensemble temperature T from the fit coefficient B, we require the value of α'/α at the same lattice trap wavelength where the carrier lineshapes are recorded. A clever way to obtain f_{ax}^2/P_{latt} is to scan the $X(62, 0) \to X(61, 0)$ Raman transition [Fig. 2.7a]. This pair of adjacently bound states have polarizability ratios very close to unity over a large range of lattice trap wavelengths ($\alpha_{61}/\alpha_{62} = 1.000932(29)$ at 914 nm [36]), permitting accurate measurement of the axial trapping frequency f_{ax} of (essentially) $X(62, 0)$ as a function of P_{latt} under diverse experimental conditions. Plugging the measured slopes L_0 and $f_{\text{ax}}^2/P_{\text{latt}}$ into Eq. (2.47) gives $\alpha'/\alpha = \alpha_0^{\text{b.g.}}/\alpha_{62} = 1.404(7)$ at 1005.6 nm; the baseline polarizability ratio of $X(0, 0)$ against $X(62, 0)$ in this case (see also Sect. 5.3).

References

1. Osborn, C.B.: The physics of ultracold Sr_2 molecules: optical production and precision measurement. PhD thesis (Columbia University, New York, 2014)
2. Sansonetti, J., Nave, G.: Wavelengths, transition probabilities, and energy levels for the spectrum of neutral strontium (Sr I). J. Phys. Chem. Ref. Data **39**, 033103 (2010)
3. Stepanov, N., Zhilinskii, B.: When and why Hund's cases arise. J. Mol. Spectrosc. **52**, 277–286 (1974)
4. Nikitin, E., Zare, R.: Correlation diagrams for Hund's coupling cases in diatomic molecules with high rotational angular momentum. Mol. Phys. **82**, 85–100 (1994)
5. Stein, A., Knöckel, H., Tiemann, E.: The ^{1}S + ^{1}S asymptote of Sr_2 studied by Fourier-transform spectroscopy. Eur. Phys. J. D **57**, 171–177 (2010)
6. Zare, R.N., Harter, W.G.: Angular Momentum: Understanding Spatial Aspects in Chemistry and Physics. Wiley-Interscience, Hoboken (1988)
7. Skomorowski, W., Pawłowski, F., Koch, C.P., Moszynski, R.: Rovibrational dynamics of the strontium molecule in the $A^1\Sigma_u^+$, $c^3\Pi_u$, and $a^3\Sigma_u^+$ manifold from state-of-the-art ab initio calculations. J. Chem. Phys. **136**, 194306 (2012)
8. Stein, A., Knöckel, H., Tiemann, E.: The states $1^1\Sigma_u^+$, $1^1\Pi_u$and $2^1\Sigma_u^+$ of Sr_2 studied by Fourier-transform spectroscopy. Eur. Phys. J. D **64**, 227–238 (2011)
9. Leung, K.H., et al.: Transition strength measurements to guide magic wavelength selection in optically trapped molecules. Phys. Rev. Lett. **125**, 153001 (2020)
10. Majewska, I.: Theoretical description of ultracold strontium molecules in an optical lattice: control of photodissociation and interpretation of molecular clock experiments. PhD thesis (University of Warsaw, Warsaw, 2021)
11. Czuchaj, E., Krośnicki, M., Stoll, H.: Valence ab initio calculation of the potential energy curves for the Sr_2 dimer. Chem. Phys. Lett. **371**, 401–409 (2003)
12. McGuyer, B.H., et al.: Precise study of asymptotic physics with subradiant ultracold molecules. Nat. Phys. **11**, 32–36 (2015)
13. McGuyer, B., et al.: Control of optical transitions with magnetic fields in weakly bound molecules. Phys. Rev. Lett. **115**, 053001 (2015)
14. Moszynski, R.: Molecular Materials with Specific Interactions–Modeling and Design, pp. 1–152. Springer, Berlin (2007)
15. Mies, F., Stevens, W., Krauss, M.: Model calculation of the electronic structure and spectroscopy of Hg_2. J. Mol. Spectrosc. **72**, 303–331 (1978)
16. Brown, J.M., Brown, J.M., Carrington, A.: Rotational Spectroscopy of Diatomic Molecules. Cambridge University Press, Cambridge (2003)
17. Herzberg, G., Mrozowski, S.: Molecular spectra and molecular structure. I. Spectra of diatomic molecules. Am. J. Phys. **19**, 390–391 (1951)
18. Zelevinsky, T., et al.: Narrow line photoassociation in an optical lattice. Phys. Rev. Lett. **96**, 203201 (2006)

19. Reinaudi, G., Osborn, C., McDonald, M., Kotochigova, S., Zelevinsky, T.: Optical production of stable ultracold $^{88}Sr_2$ molecules. Phys. Rev. Lett. **109**, 115303 (2012)
20. McGuyer, B., et al.: High-precision spectroscopy of ultracold molecules in an optical lattice. New J. Phys. **17**, 055004 (2015)
21. McDonald, M.: High Precision Optical Spectroscopy and Quantum State Selected Photodissociation of Ultracold $^{88}Sr_2$ Molecules in an Optical Lattice. Springer, Berlin (2017)
22. Lee, C.-H.: Quantum metrology with a molecular lattice clock and state-selected photodissociation of ultracold molecules. PhD thesis (Columbia University, New York, 2020)
23. McDonald, M., et al.: Photodissociation of ultracold diatomic strontium molecules with quantum state control. Nature **535**, 122–126 (2016)
24. McDonald, M., et al.: Control of ultracold photodissociation with magnetic fields. Phys. Rev. Lett. **120**, 033201 (2018)
25. Kondov, S., et al.: Crossover from the ultracold to the quasiclassical regime in state-selected photodissociation. Phys. Rev. Lett. **121**, 143401 (2018)
26. Majewska, I., et al.: Experimental and theoretical investigation of the crossover from the ultracold to the quasiclassical regime of photodissociation. Phys. Rev. A **98**, 043404 (2018)
27. Grimm, R., Weidemüller, M., Ovchinnikov, Y.B.: Optical dipole traps for neutral atoms. Adv. At. Mol. Opt. Phys. **45**, 95–170 (2000)
28. Hilborn, R.C.: Einstein coefficients, cross sections, f values, dipole moments, and all that. Am. J. Phys. **50**, 982–986 (1982)
29. Aeppli, A., et al.: Hamiltonian engineering of spin-orbit–coupled fermions in a Wannier-Stark optical lattice clock. Sci. Adv. **8**, eadc9242 (2022)
30. Kolkowitz, S., et al.: Spin–orbit-coupled fermions in an optical lattice clock. Nature **542**, 66–70 (2017)
31. Leibfried, D., Blatt, R., Monroe, C., Wineland, D.: Quantum dynamics of single trapped ions. Rev. Mod. Phys. **75**, 281 (2003)
32. McDonald, M., McGuyer, B.H., Iwata, G.Z., Zelevinsky, T.: Thermometry via light shifts in optical lattices. Phys. Rev. Lett. **114**, 023001 (2015)
33. Blatt, S., et al.: Rabi spectroscopy and excitation inhomogeneity in a one-dimensional optical lattice clock. Phys. Rev. A **80**, 052703 (2009)
34. Blatt, S.: Ultracold collisions and fundamental physics with strontium. PhD thesis (University of Colorado at Boulder, Boulder, 2011)
35. Han, C., et al.: Carrier thermometry of cold ytterbium atoms in an optical lattice clock. Sci. Rep. **8**, 1–8 (2018)
36. Leung, K.H., et al.: Ultracold $^{88}Sr_2$ molecules in the absolute ground state. New J. Phys. **23**, 115002 (2021)

Chapter 3
Frequency Comb Assisted Spectroscopy of the States $X^1\Sigma_g^+$, $(1)0_u^+$, and $(1)1_u$

3.1 Spectroscopy Laser System

3.1.1 Master Laser, and the Pound-Drever-Hall Technique

For the majority of experiments in this thesis, an extended cavity diode laser (ECDL) acts as the master laser; i.e., a metrological grade laser with the best optical coherence in the lab, serving as a stable phase reference for other important lasers. This exceptional phase stability is realized by stabilizing the ECDL using the **Pound-Drever-Hall** (PDH) technique to a high finesse reference cavity[1] formed by two fused-silica mirrors[2] optically bonded to an ultra-low expansion (ULE) glass spacer[3]. The ULE spacer defines the cavity length (approximately 10 cm), resulting in a free spectral range (FSR) of approximately 1.497 GHz, measured by locking the master laser to adjacent cavity modes and determining the change in laser frequency using the frequency comb (see next subsection). Laser light enters the cavity at the plano-plano mirror, while the other mirror is plano-concave with 50 cm radius of curvature. The back surfaces of both mirrors are applied with anti-reflective coating to minimize etalon effects. The finesse ($\mathcal{F}$) is $> 3 \times 10^5$ around 780 nm (measured from ring-down experiments [2]). In principle, this implies a cavity linewidth of $\mathrm{FSR}/\mathcal{F} \lesssim 5$ kHz. When the laser is on resonance with a cavity mode, the intracavity power is larger than the input power by a factor $\sim \mathcal{F}$. Empirically, the resonance frequency of the cavity is dependant on the input laser power [3]. As such, care

[1] Full commercial system including vacuum housing purchased from Stable Laser Systems.

[2] Fused-silica substrates have lower thermal noise floors than ULE. Both fused-silica mirrors are 1" in diameter and 0.25" in thickness, and each is optically contacted with a 1" ULE backing ring to reduce thermally induced mirror deformation [1].

[3] ATF 6020-4 notched cavity. Notched cavities are less sensitive to vibrations than cylindrical cavities.

K. H. Leung, *The Strontium Molecular Lattice Clock*, Springer Theses,
https://doi.org/10.1007/978-3-031-47647-1_3

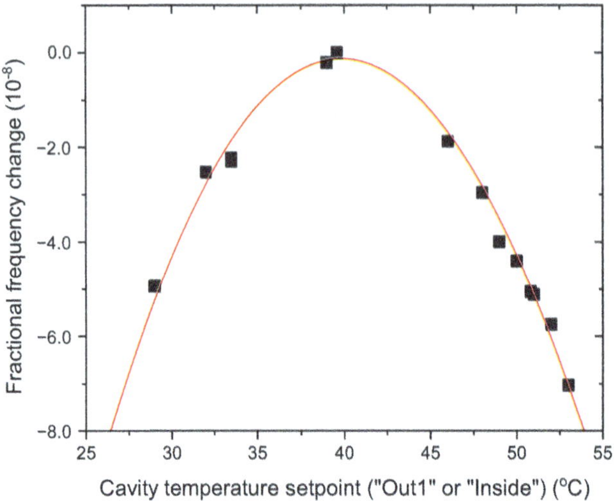

Fig. 3.1 Fractional change of the master laser (relative to an arbitrary value) versus the cavity temperature setpoint (corresponding to "Out1" or "Inside" on the PTC10 temperature controller). The coefficient of thermal expansion has a zero crossing around 39.6(3) °C. Recorded during December 2019

should be taken not to use excessive laser power under lock conditions.[4] The cavity rests horizontally relative to gravity, and the structure is supported at judiciously selected stress points by Viton balls on a Zerodur block.

The cavity and its supporting structure are placed under vacuum conditions ($< 10^{-7}$ Torr). The vacuum housing windows are anti-reflection coated for light at 780 nm (specified < 0.2% reflection coefficient for normal incidence), again to prevent spurious etalons. The housing offers two-stage temperature control, and we set the cavity temperature (τ) at the zero crossing temperature (τ_c) of the ULE spacer's coefficient of thermal expansion. We determine τ_c by locking the repetition rate of an optical frequency comb to the master laser (see next subsection). The absolute frequency of the master laser (f_{laser}) was measured at various τ, as shown in Fig. 3.1. τ was changed at 24-hour intervals so that the cavity and its neighboring structure are given enough time to equilibrate. The fractional change in the laser frequency ($\Delta f_{\text{laser}}/f_{\text{laser}}$) has a turning point at τ_c, where the sensitivity to temperature fluctuations is minimal [4],

$$\frac{\Delta f_{\text{laser}}}{f_{\text{laser}}} = -\frac{a}{2}(\tau - \tau_c)^2 - \frac{b}{3}(\tau - \tau_c)^3 + \ldots . \tag{3.1}$$

We determine $\tau_c = 39.6(3)$ °C. Lastly, the entire vacuum housing and optical (and optoelectronic) elements for PDH stabilization are mounted on a passive vibration isolation platform (Minus K 250BM-1). The floating platform is completely covered

[4] Typical input laser powers into the cavity under locking conditions are ~ 20–60 μW.

by an air-tight enclosure. The single-mode polarization maintaining fiber delivering light from the master laser into the enclosure is short (20 cm) and surrounded by a 1" thick flexible rubber foam pipe tube to passively reduce fiber phase noise due to perturbations from the ambient environment. Under these conditions, consolidated daily measurements of the absolute frequency of the master laser over several months reveal a linear[5] cavity drift rate of 0.03 Hz/s. This linear drift is removed with real-time feedforwarding using an AOM in the optical path between the master laser and the reference cavity. The frequency synthesizer that performs this linear compensation updates every second.

PDH laser stabilization to an optical cavity is a powerful and widely used technique to achieve Hz-level laser linewidths. For a useful technical reference, see Ref. [5]. Essentially, the cavity stores previously resonant laser light, the leakage of which is compared with the present state of the laser to generate an error signal. In an ideal situation with no losses, when laser light is exactly on resonance with a cavity mode, all the power of the incident laser beam will be transmitted through the cavity, and no power is reflected. This is because the total reflected field is the superposition of the so-called *promptly-reflected* field (the first reflection at the incident mirror) and the so-called *leakage* field (the coherent sum of multiple reflections inside the cavity that emanates out of the incident mirror). On resonance, the promptly-reflected and the leakage fields have the same frequency and are π-radians out of phase, resulting in complete destructive interference. However, when the laser frequency is slightly off-resonance, the resulting phase difference leads to a small amount of laser power being reflected from the cavity. The phase and amplitude of this reflected field encode the sign and magnitude of the frequency deviation with respect to the cavity mode, respectively. Detection of this field can thus be used to actively servo the laser back onto resonance with the cavity. In practice, the phase of an optical field is difficult to detect in real-time. Instead, the frequency or phase of the laser is modulated (i.e., dithered) to sample the derivative of the cavity mode profile, thereby determining which side of the resonance the laser frequency is on.

A scheme of the optical (and optoelectronic) elements is shown in Fig. 3.2. An optical isolator is installed in the optical path prior to the cavity to suppress etalon effects, which would otherwise result in a non-uniform baseline in the error signal. Note while an additional photodiode for intensity stabilization has been installed, for the experiments in this thesis, the incident laser intensity at the cavity is not actively stabilized to keep the apparatus simple; the free-running intensity drift is fewer than 5%. In our setup, a weakly driven resonant electro-optic modulator (EOM) creates frequency[6] modulation (FM) sidebands on the laser. This frequency is usually large (e.g. 30 MHz) to provide immunity to external RF interference in the

[5] Note that an exponentially decaying drift rate would better model the aging of the cavity over years.

[6] Strictly speaking, an EOM creates a phase modulation, but for the present discussion, we may ignore this distinction.

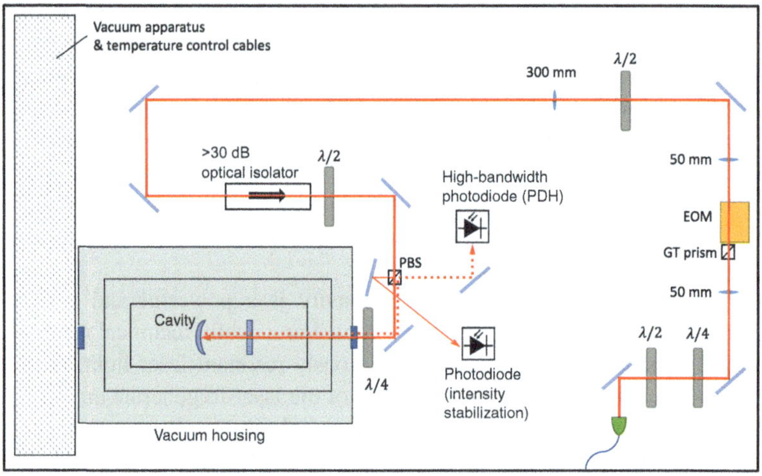

Fig. 3.2 Cartoon illustration of the optical setup for PDH locking to the high-finesse cavity

lower frequency bands. Let us consider the limit where the modulation frequency is much larger than the cavity mode linewidth, as is typically the case for high-finesse cavities. When the carrier is on resonance with a cavity mode, it is fully transmitted. The sidebands, however, are reflected toward the photodiode. Now consider the situation when the carrier is slightly detuned from a cavity mode. A small percentage of the carrier also makes it back to the photodiode, and a beat with the sidebands at the EOM modulation frequency is detected. The phase of this beat, which contains the phase of the reflected field, can be detected with conventional RF electronics because it is a microwave frequency. This is done by demodulating the beat signal with an RF mixer or a phase detector, thereby creating the PDH error signal [Fig. 3.3a]. Note that the discriminators of the sidebands have the opposite sign from that of the carrier, and it is possible to erroneously lock these sidebands to the cavity if the sign and input offset on the servo is set incorrectly.[7] The cavity transmission when the carrier laser frequency is locked to the lowest order (TEM00) Gaussian mode of the cavity is always the brightest, most intense. A quantitative account of the PDH error signal and its generation is given in Ref. [6]. The error signal is subsequently fed to a high-bandwidth analog servo controller (Toptica FALC pro), which actuates directly on the current driving the ECDL laser diode to tune its emission frequency. In this manner, fast phase fluctuations of the laser relative to the cavity resonance are actively nulled. The servo also provides a slower

[7] For less demanding, non-metrological applications, FM sidebands present a feature, rather than a complication. A rather common experimental trick to deterministically sweep a laser across a large frequency range involves using a broadband non-resonant fiber EOM to generate large FM sidebands (~0.1–10 GHz) on top of the smaller FM sidebands (~10 MHz) used for PDH locking to a cavity or a gaseous atomic reference cell.

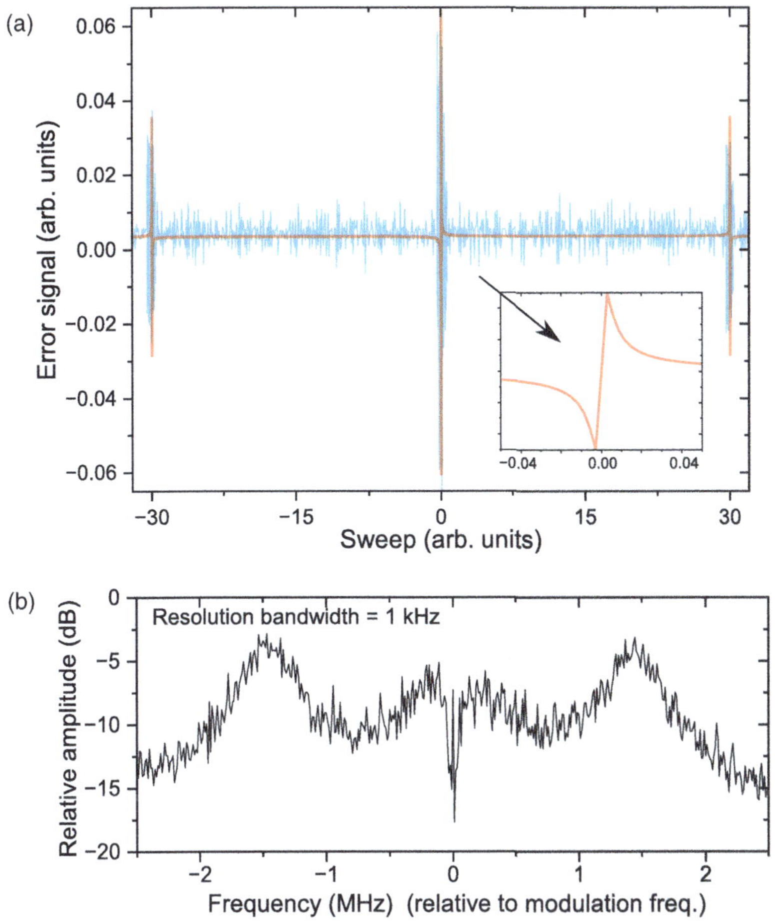

Fig. 3.3 (**a**) PDH error signal for locking the ECDL to the high finesse cavity, obtained by sweeping the laser across the cavity mode and recording the in-loop photodiode signal on an oscilloscope triggered to the sweep. As the cavity linewidth is extremely narrow and the free-running laser frequency drifts very quickly, it is difficult to achieve clean traces. Blue trace, data; red trace, simulated error signal for $\mathcal{F} = 3 \times 10^5$. Inset: Magnified portion of the carrier error signal discriminator. (**b**) Noise spectrum measured by the in-loop photodiode with the lock is engaged. The servo bumps are exaggerated here for visual clarity (their heights are slightly lower for normal gain settings). The locking bandwidth is about 1.5 MHz. The input offset on the servo is adjusted to minimize the central spike (note that this action modifies the frequency lock point of the laser relative to the cavity mode)

auxiliary output that is used to actuate on a piezo element on the ECDL to correct for longer-term frequency drifts. Figure 3.3b shows an example noise spectrum of the in-loop photodiode (beat) signal centered at the EOM modulation frequency when the lock is engaged, obtained by picking off a small portion of the signal with an RF coupler.

A general rule of thumb is that all electronic or RF components (e.g., the servo controller, current controller, photodiode, length of cables, etc.) should have at least an order-of-magnitude larger bandwidth than the desired locking bandwidth. To cut possible ground loops, we use RF baluns[8] at the mixer inputs, and cable lengths are kept as short as possible. The stabilized laser has a high degree of optical phase stability. The laser linewidth can only be rigorously determined by comparison with another laser of similar performance (via a low noise frequency comb as a transfer oscillator if necessary). Unfortunately, this is not currently feasible in the lab.[9] Nevertheless, from analyses of the stability of molecular clock spectra (see Chap. 5), it is quite likely the cavity linewidth is on the order of a few Hz, and definitely smaller than 100 Hz.

In principle, the PDH error signal is immune to fluctuations in the laser intensity because of nulled lock-in detection (i.e., on resonance, the laser intensity is zero at the photodiode that registers the beat note). However, imperfections in a realistic setup can contaminate the error signal and degrade the achievable lock performance. One chronic limitation of EOM phase modulation is the concomitant **residual amplitude modulation** (RAM), which also oscillates at the modulation frequency. The servo erroneously reacts to RAM because it is indistinguishable from the cavity response, writing noise onto the laser phase. Detailed quantitative analyses on the effects of RAM, specifically in PDH laser stabilization schemes, are available in literature [8]. The temperature-dependent birefringence of the EO crystal has been identified as a leading source of RAM.[10] If the input laser beam is not perfectly linearly polarized parallel to the extraordinary axis of the EO crystal, the crystal acts as a variable waveplate, and the output polarization is perturbed at the rate of the modulation frequency. Subsequent transmission through a polarizing element, such as a polarizing beam splitter (PBS), leads to amplitude modulation. As one might imagine, perfect polarization alignment to a time-varying axis is challenging to achieve in realistic situations.

While well-established methods exist to actively cancel or stabilize RAM [9, 10], these approaches generally utilize nearly dc electric fields applied to a non-resonant EOM, which necessitates large voltages. In the current version of our setup, we opted for an off-the-shelf free-space resonant EOM as it can be easily driven by standard RF synthesizers.[11] Therefore, we undertake passive strategies to reduce the effect of RAM. As the birefringence is temperature dependent, we thermally contact the EOM outer housing with a copper block to increase the thermal mass of

[8] Minicircuits FTB-1-1*A15+.

[9] The other cavity which stabilizes 689 nm strontium cooling light has a linewidth of about 200 Hz [7].

[10] Other sources include etalon effects between the surfaces of the EO crystal or other optical elements, scattering effects, and spatial inhomogeneity of the laser beam sampling different regions of the EO crystal. To mitigate the latter, the beam should not be clipped (e.g., by irises or optical elements), and ideally should not be focused onto the photodiode.

[11] Essentially, a simplistic view of a resonant EOM is that it has an RLC tank circuit which amplifies the input ac electric field on resonance.

the crystal's immediate environment. The EOM is fixed on a 5-axis stage which is mounted on a ceramic post to insulate the structure from the colder optical breadboard. The air-tight box around the entire optical setup also further helps mitigate convective heat exchange from the hustle and bustle of the lab. The next step involves careful alignment of the laser polarization to the crystal's extraordinary axis. To do so, we monitor the signal from the photodiode for PDH locking on a spectrum analyzer centered at the EOM modulation frequency. Importantly, the laser is tuned far away from a cavity mode (ideally in the middle of two cavity modes) to decouple the photodiode signal from the cavity response. The RAM signal appears as a small narrow peak at the modulation frequency and may be verified by turning the modulation off or blocking the laser beam. With the Glan-Thompson (GT) prism initially removed, a pair of half and quarter waveplates (in series before the EOM) are rotated iteratively to minimize the RAM peak on the spectrum analyzer, while simultaneously maintaining maximum incident power on the photodiode (e.g., using the half waveplate before the optical isolator). Once an optimal setting is found, the GT prism with a high extinction ratio (10^5:1) is installed between the waveplates and EOM, and oriented to maximize the transmitted beam through the EOM [Fig. 3.2]. Slight adjustments on the 5-axis stage on which the EOM is mounted and sequential re-optimization of the cavity alignment using the penultimate and final mirrors may be necessary. Again, as the orientation of the extraordinary axis is temperature dependent, this entire process is repeated whenever the ambient conditions of the lab change throughout the seasons, or before a clock frequency measurement campaign. While this passive strategy is sufficient for this thesis, as the demand for greater stability inevitably grows, we anticipate that active RAM servo will become necessary for future work.

3.1.2 Optical Frequency Comb, and the Lab Time Base

The optical frequency comb is one of the most significant inventions of the past 20 years. Developed at the turn of the new millennium, frequency combs establish a direct experimental link between optical and microwave frequencies, and are vastly simpler to operate than archaic and unwieldy frequency chains. Among the known physical quantities, the phase accumulation of a wave can be determined with exceptional accuracy because it is measured solely using the method of counting, which has no fundamental bound on its uncertainty. Conceptually, the value of a physical measurement given in terms of a standardized unit is just the ratio of two integers, of which the denominator is set and defined. Therefore, the accuracy of a physical measurement fundamentally depends on how closely we can realize the idealized definition of the unit with real physical apparatus. As the progenitor quantum device, atomic clocks are unparalleled in their level of fractional stability, accuracy, reproducibility, and transportability. For this reason, in the current definition of the SI system of units, the second (unit of time) remains the only quantity referencing a physical object (the hyperfine clock transition in the cesium atom). In contrast, all other SI units (except for the mole) are tethered to the SI second and defined by the fundamental constants of nature; i.e., speed of light, Planck constant, elementary charge, Boltzmann constant, Avogadro constant, luminous efficacy of 540 THz green light. As discussed in Chap. 1, the larger the operating frequency of a clock, the more "ticks" (phase cycles) are available for counting in a given length of interrogation time, and the easier it becomes to make a highly stable and accurate clock. Optical frequency combs provide the crucial piece of clockwork for frequency metrology above microwave frequencies by enabling accurate measurements of optical-optical and optical-microwave frequency ratios.

Although other techniques also exist, the archetypal frequency comb is generated by a mode-locked pulsed laser. In our lab, the frequency comb is a piece of commercial equipment based on a femtosecond erbium-doped fiber laser.[12] The central idea is that a train of pulses emitted by a mode-locked laser at repeated intervals (Δt_{RR}) is a comb of evenly spaced spectral lines in the frequency-domain [Fig. 3.4a and b]. Due to dispersion in the laser cavity, there is a phase offset ($\Delta\phi_{\mathrm{CEO}}$) between the carrier and the envelope that accumulates from pulse-to-pulse. This phase offset introduces a frequency offset $f_{\mathrm{CEO}} \simeq 1/(\Delta t_{\mathrm{CEO}})$ to the comb teeth, called the **carrier-envelope offset**, such that they are not exact harmonics of the **repetition rate** $f_{\mathrm{RR}} \simeq 1/(\Delta t_{\mathrm{RR}})$. The frequency of the N-th comb tooth is given by the comb equation [11, 12]

$$f_{\mathrm{comb}}^{(1\times)} = N f_{\mathrm{RR}} + f_{\mathrm{CEO}}. \tag{3.2}$$

Note that the **sign** of f_{CEO} may be positive or negative as it is technically a beat between the comb with itself [Eq. (3.4)]. Each comb tooth may be considered a continuous wave laser, with a well-defined phase relationship with the other comb teeth. If Δt_{env} is the duration of the pulse envelope, then the resulting comb will be spread around the carrier frequency of the mode-locked laser with width $\sim 1/(\Delta t_{\mathrm{env}})$. For our comb, this fundamental output[13] is centered around 1560 nm with a full width at half maximum (FWHM; −3 dB point) of about 40 nm. This output is amplified by a fiber amplifier and made octave spanning using a highly non-linear fiber,[14] which broadens the spectrum to the range ~1–2 μm. Sum frequency generation (SFG) in a non-linear crystal not only shifts the central comb carrier, but also results in a comb spectrum that has a zeroth comb tooth at **twice** the frequency of f_{CEO},

[12] Menlo FC1500-250-ULN, with custom add-ons. Upgraded from an older model that we had before 2021.

[13] The M-Comb module (femtosecond, mode-locked, erbium-doped fiber ring laser) has 6 of these comb outputs at 1560 nm. One of which is used to extract f_{CEO} (with the f–$2f$ interferometer inside the P250-XPS-WG module). At the time of writing, 3 out of the remaining 5 outputs are used to generate comb light at our operational wavelengths through the various add-ons: M-NIR, M-VIS, and HMP780+760 (these are separate ports).

[14] Note that this is the same process that produces the M-NIR output.

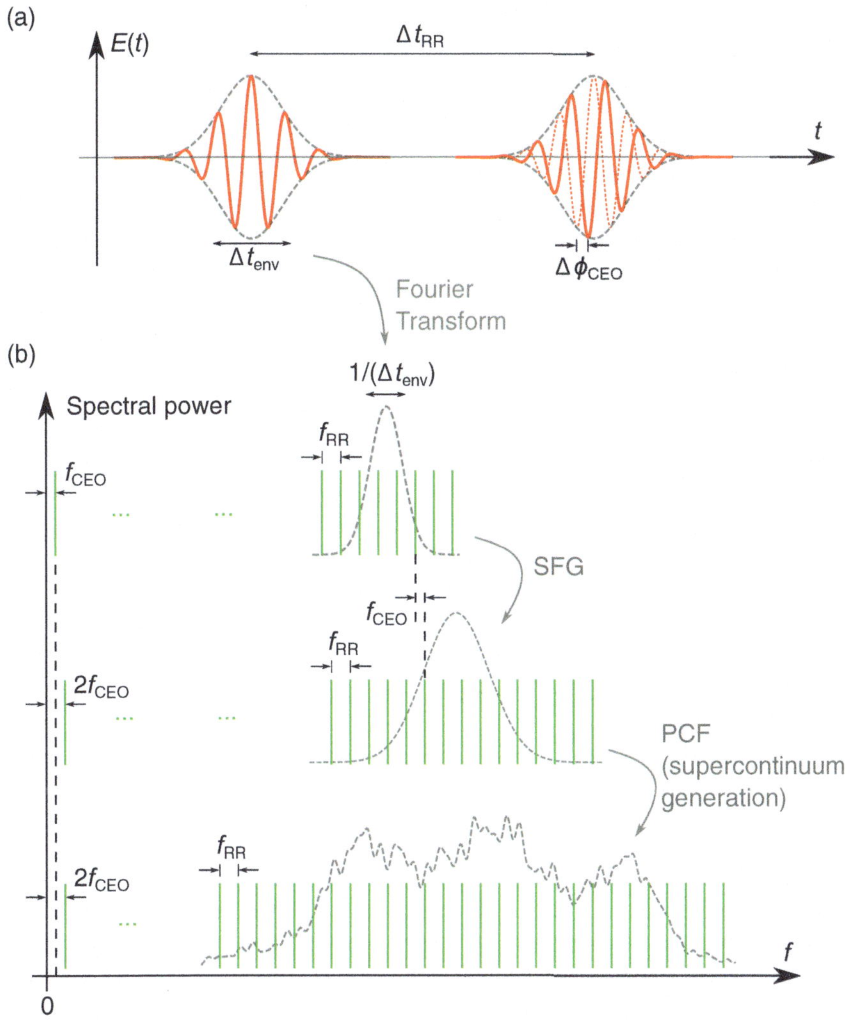

Fig. 3.4 (**a**) Electric field as a function of time of a pulsed mode-locked laser; carrier wave (red), pulse envelope (dotted gray). (**b**) The corresponding frequency comb spectrum in Fourier space (green lines). Comb envelope functions after various non-linear processes are illustrated by the dotted gray cartoon curves. Starting from the fundamental comb spectrum centered around the laser carrier (1560 nm), a supercontinuum can be generated. Note that the actual f–$2f$ interferometry involves broadened comb light in a highly-nonlinear fiber ranging from 1 to 2 μm, unlike what is shown. The number of comb teeth and their spacing are not drawn to actual scale

$$f_{\text{comb}}^{(2\times)} = N f_{\text{RR}} + 2 f_{\text{CEO}}. \tag{3.3}$$

This offers a way to extract f_{CEO} and self-reference the comb using a so-called f–$2f$ interferometer. Performing sum frequency generation to the $\sim 2\,\mu$m part of the broadened comb and beating it with the undoubled $\sim 1\ \mu$m part on a photodiode, the lowest measured frequency will be

$$f_{\rm CEO} = f^{(2\times)}_{\rm comb} - f^{(1\times)}_{\rm comb}. \tag{3.4}$$

Higher harmonics at $f_{\rm RR} - f_{\rm CEO}$, $f_{\rm RR}$, $f_{\rm RR} + f_{\rm CEO}, \ldots$, are rejected with a lowpass filter. Importantly, the sign of $f_{\rm CEO}$ is defined by Eq. (3.4); i.e., if $f^{(2\times)}_{\rm comb} - f^{(1\times)}_{\rm comb}$ is positive (negative), then we say that the carrier-envelope offset is positive (negative). A practical method of determining the carrier-envelope offset sign is outlined below, toward the end of this subsection. Typically, the signal-to-noise of $f_{\rm CEO}$ at the corresponding locking electronics on the comb rack is >40 dB in a 100 kHz resolution bandwidth.

For spectroscopy of Sr_2 in this thesis, we typically use lasers in the visible and near-infrared (NIR). We already saw that a highly non-linear fiber can broaden the fundamental comb to cover the NIR. To generate comb light in the visible (e.g. $<$ 800 nm), we first amplify the fundamental comb output around 1560 nm and perform SFG in a non-linear crystal (periodically-poled lithium niobate, PPLN), producing a narrow band of comb light centered around 780 nm.[15] To further broaden this to useful visible wavelengths, the doubled comb is coupled[16] into a specially designed photonic crystal fiber (PCF) where tiny air holes parallel to the fiber core modify the refractive index, inducing a frequency chirp on the time-domain laser pulses. The result is a supercontinuum[17] of comb light in the range $\sim$ 600–1000 nm.

To detect the beat between a laser and the frequency comb, we implement the free space[18] scheme in Fig. 3.5a. The laser and comb light are coaligned on polarizing beam splitters (PBS). A diffraction grating with high resolvance (i.e., having a high number of grating lines per unit length, and oriented so that as many grating lines are illuminated by the laser beams) helps separate the redundant portion of the comb spectrum from the relevant laser wavelength, so that only comb teeth in the vicinity of the laser wavelength is incident on the photodiode. This is vital to acquire a beat note with a high signal-to-noise ratio. In practice, the power ratio of the laser and comb light is also optimized (using the waveplate at the PBS just before the grating)

---

[15] This process produces the visible HMP (High Power Measuring Port) outputs. Because the spectral power is concentrated into fewer comb teeth, beats with continuous wave lasers can be very strong (>33 dB, usually almost 40 dB9 in a 100 kHz resolution bandwidth).

[16] This fiber coupling should be occasionally optimized (mirror beam walk) to keep the M-VIS power high (>90 mW measured immediately after the PCF with a power meter at a wavelength setting of 780 nm).

[17] This is the M-VIS output. As the spectral power is spread out over a larger number of comb teeth, it is often necessary to optimize the diode current values of the fiber amplifier (controlled by the corresponding channel on the AC1550) to maximize the comb light at the desired wavelength to acquire strong beats (>33 dB in a 100 kHz resolution bandwidth).

[18] Free space beat detection has the advantage of not accruing fiber phase noise, an important source of instability when performing clock spectroscopy.

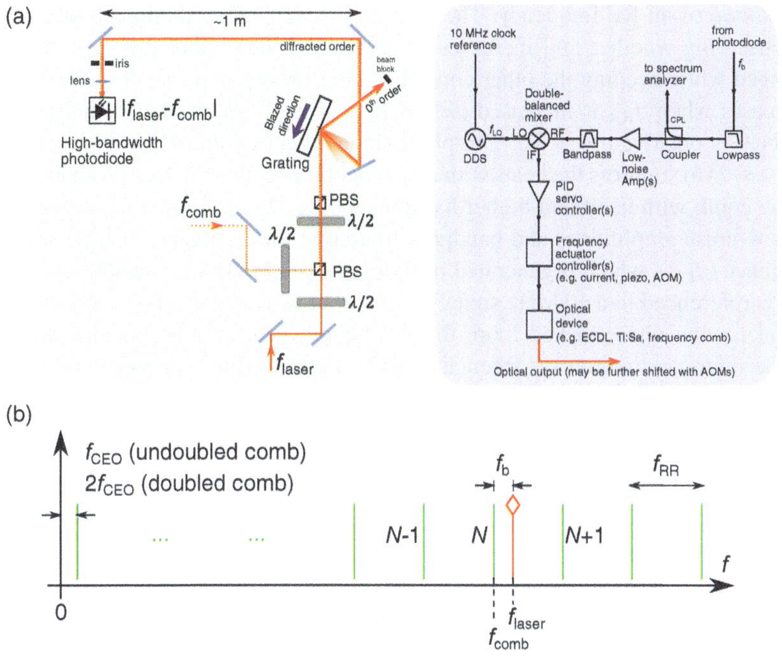

Fig. 3.5 (**a**) Optical (left) and RF electronic (right) setups for a phase lock between a laser and the frequency comb. In the free space beat detection unit, the laser and the comb light are coaligned and diffracted by a grating with high resolvance. The diffracted order is made to spread in space to separate out the wavelengths so that only comb teeth around the relevant laser wavelength are incident on a high bandwidth photodiode. The photodiode detects the beat of the laser and the comb, among other frequencies suppressed by a low pass and band pass filter in the RF electronic chain. (**b**) Laser-comb beat diagram. We typically utilize positive beats; i.e., the frequency of the laser (vertical red line with diamond) is greater than that of the nearest comb tooth (green lines). We reiterate that the magnitude of the frequency of the zeroth comb tooth depends on whether the comb light derives from the fundamental (i.e, undoubled) output or from comb light that has undergone sum-frequency generation. The number of comb teeth and their spacing are not drawn to scale

to maximize the signal-to-noise ratio of the beat. The incident polarization on the grating should maximize the diffraction efficiency at the desired wavelength (i.e., s-polarization versus p-polarization). An iris is placed before the lens to further block the redundant comb spectrum before a lens focuses the beams onto a high-bandwidth photodiode.

The photodiode outputs the difference (or beat) frequency of the laser light and the nearest comb tooth (with tooth number N),

$$f_b \equiv f_{laser} - f_{comb}^{(s\times)} \tag{3.5}$$

By convention, we say that the **beat is positive if** $f_b > 0$, **and negative if** $f_b < 0$; see also Fig. 3.5b. Spectrum analyzers and frequency counters only ever measure the *magnitude* of an RF frequency. The sign of f_b can be unambiguously determined by observing whether $|f_b|$ increases or decreases when either the f_{laser} or f_{RR} is changed while keeping the other constant.[19] For example, if one observes that $|f_b|$ decreases when f_{RR} is increased (while f_{laser} is kept essentially unchanged), then the beat is positive ($f_b > 0$). The photodiode signal is filtered by a lowpass filter ($< f_{RR}/2$) to suppress the beats with adjacent comb teeth ($N \pm 1$), as well as the beat of the comb with itself and higher harmonics ($f_{RR}, 2f_{RR}, \ldots$). It is then amplified by low-noise amplifier(s) and bandpass filtered at the frequency of the microwave reference (f_{LO}), which is generated by a direct digital frequency synthesizer (DDS) that is referenced to a 10 MHz signal from a microwave clock. To generate an error signal for the phase-locked loop, the laser-comb beat (f_b) is mixed down to dc by the reference microwave frequency. This is then fed to a proportional-integral-differential (PID) servo controller tuned to maximize the overall locking bandwidth. When the lock is engaged, the servo controls the frequency actuator on the optical device (either the comb or the laser) to ensure that $f_b = f_{LO}$. Note that robust mixing requires that the power supplied to the LO port of the mixer matches the specified mixer level. For example, an LO power of +3 dBm should be used for a level 3 mixer. The mixer's IF port should be as close as possible to the servo controller, ideally directly connected with coaxial adapters or a very short coaxial cable. The response of the PID servo naturally filters out high frequencies in the error signal, and a lowpass filter at the IF output port is generally not needed. Note that while input dc offsets on phase-locked loop error signals do not affect the frequency lock point, most stable locks tend to have dc offsets close to zero as it maximizes the lock capture range.

To **determine the sign of the carrier-envelope offset**, f_{CEO} and f_{RR} are first tuned by hand to their operating values. While leaving f_{CEO} free running, the servo on f_{RR} is engaged (see below) to phase lock the nearest comb tooth to a stable laser (e.g. the master laser described in the previous subsection). This guarantees that the tooth number N and the beat frequency are fixed, $f_{LO} = f_b = f_{laser} - f_{comb}^{(s\times)} = f_{laser} - Nf_{RR} - sf_{CEO}$, where the multiplicative constant $s = 1$ or 2 because the frequency of the zeroth comb tooth depends on whether the comb light has undergone sum-frequency generation or not; see Eqs. (3.2) and (3.3). Rearranging, we find that $f_{RR} = (f_{laser} - f_b - sf_{CEO})/N$. Note that f_{RR} is always positive because, by definition, $N \geq 0$ and the frequency of an electromagnetic wave is positive. Next, we change $|f_{CEO}|$ by $\Delta|f_{CEO}|$ to observe the corresponding change in f_{RR} on a frequency counter with sub-Hz level resolution (a spectrum analyzer is not sensitive enough to resolve the change). For example, if the carrier-envelope offset is positive, increasing $|f_{CEO}|$ will decrease f_{RR} by a magnitude $s\Delta|f_{CEO}|/N$. In this manner, the sign of the carrier-envelope offset is

[19] One can tell the direction of the change in f_{laser} on a wavemeter, and that of f_{RR} on a frequency counter.

unambiguously determined.[20] As verification of this procedure and a sanity check, we measure the absolute frequency of the $^1S_0 \rightarrow {}^3P_1$ intercombination transition of atomic ^{88}Sr in free space, finding excellent agreement with the published literature values.[21]

For the experiments in this thesis, a microwave rubidium clock[22] serves as the **laboratory timebase**. Its "10 MHz" clock outputs are amplified and split while preserving the exquisitely low phase noise by a 14-channel distribution amplifier.[23] All important DDS (including those on the frequency comb rack), various function generators, and frequency counters are referenced to the +10 dBm outputs of the distribution amplifier. Many of these synthesizers produce microwave sine waves that act as reference frequencies for phase-locked loops or to drive AOMs that diffract probe laser beams, so it is essential that the RF frequencies can be traced back to the laboratory timebase. Rubidium microwave clocks are known to drift over durations of >1 day [13]; i.e., the "10 MHz" rubidium clock output may not actually be 10 MHz according to a more accurate standard. Over long periods, the random walk can accumulate rather large frequency offsets. To alleviate this, for day-to-day operations, we discipline the rubidium clock to the 1 pulse-per-second (PPS) signal from a GPS disciplined oscillator (GPSDO)[24] with a time constant of approximately 18 hours. This eliminates the long-term drift of the rubidium clock at the expense of discretized jumps in frequency (few parts in 10^{12}) every few hours due to its settability. To obtain a calibration of the rubidium clock below 10^{-12} for demanding applications (e.g., clock frequency measurements at the 13-digit level), we instead leave the rubidium clock free-running (i.e., undisciplined and disconnected from the GPSDO) and measure its average frequency against a dual-band global navigation satellite system (GNSS) receiver[25] on a time interval counter. In post-process, the fractional frequency offset of the rubidium clock is then propagated to "post-correct" all synthesizers and frequency counters that have been referenced to it; see Sect. 5.7 for full details, as well as a note of how GPS time relates to International Atomic Time (TAI) and the SI system of units. Future

[20] Note that specious conclusions will be drawn had we also left the repetition rate free running. This is because actuating f_{CEO} also slightly actuates f_{RR}. This gets multiplied by the large comb tooth number N, leading to a substantial change in the beat frequency $f_{\mathrm{laser}} - f_{\mathrm{comb}}$ that can mask the sign of Δf_{CEO}.

[21] The intercombination wavelength is approximately 689 nm. Naturally, this means using the M-VIS comb spectrum, whose zeroth comb tooth is at $2 f_{\mathrm{CEO}}$. Therefore, measuring the intercombination not only checked the sign of the carrier-envelope offset but also verified our understanding of the doubling effect that sum-frequency generation has on the carrier envelope offset.

[22] SRS FS725.

[23] SRS FS735/11.

[24] Connor-Winfield FTS500. GPS L1 signal received by an HP 58532A antenna on the roof of the building and transferred to the lab with the RavenLink RVL-1 optical fiber setup.

[25] SparkFun ZED-F9T. A Tallysman TW3972XF antenna receives GPS L1 and L2C signals on the roof of the building and is transferred to the lab directly via a coaxial cable.

experiments should benefit from a microwave standard with better instability (e.g., a cesium beam tube standard or a hydrogen maser).

f_{laser}, the absolute frequency of the laser light *where it becomes coaligned with the comb light*, can thus be fully determined in SI units given knowledge of f_{b} (equal to the synthesized microwave frequency f_{LO} when the phase-locked loop is engaged), N (rounded to the nearest integer from initial estimates using a wavemeter), f_{RR} (using a frequency counter), f_{CEO} (using a frequency counter) and its multiplicative factor $s \in \{1, 2\}$, and the corresponding *signs* of f_{b} and f_{CEO}. The frequency of the laser light at the molecules may be different compared to where it is coaligned with the comb because of frequency shifters (e.g. AOMs) in the optical paths. These must be carefully accounted for when determining molecular and atomic transition frequencies.

Figure 3.6 shows the typical overlapping Allan deviation, $\sigma_y(\tau)$, of f_{CEO} and f_{RR}, computed from a time series of the relevant frequencies logged by a frequency counter[26] with 1 s gate time and referenced to the lab timebase. Here, the rubidium timebase was left free running, or otherwise, its discretized settability would set a $\sim 10^{-12}$ floor, masking the better short-term instability of f_{RR} (referenced to the high-finesse cavity). To stabilize f_{CEO}, we phase lock f_{CEO} to a synthesized

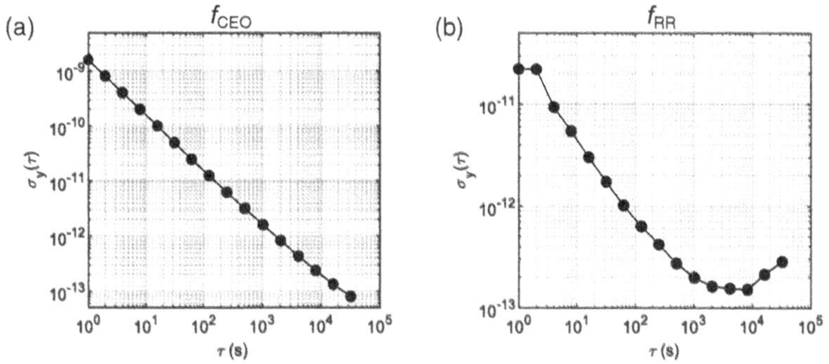

Fig. 3.6 Overlapping Allan deviation computed for the (**a**) carrier-envelope offset and the (**b**) repetition rate of the optical frequency comb under typical lock conditions. The frequencies are counted against a free running rubidium timebase

[26] K+K FXM50, a zero dead time Π-type counter. At the time of this writing, our lab has recently procured the updated K+K FXE model, which can be set to operate in either Π- or Λ-type counting modes, also with zero dead time. Λ-type counters afford lower measurement noise by internally averaging the input signal. However, as a note of caution, extra attention should be paid to the Allan deviation estimator computed from a time series recorded by Λ-type counters, as the estimator may become distorted or modified [14–16]. While there are no strict rules, a quick survey of literature in our field reveals that metrology groups tend to utilize counters operated in the traditional Π-type mode to count beat frequencies for absolute frequency measurements, even if it may not be optimal, perhaps for the sake of clarity and interpretation.

microwave frequency at 20 MHz, and $f_{\rm CEO}$ is directly counted. To optically stabilize $f_{\rm RR}$, we phase lock the beat frequency between the master laser and the nearest comb tooth to another synthesized microwave frequency at 20 MHz. To count $f_{\rm RR}$, we utilize a common trick involving higher harmonics—we mix the 4th *harmonic* of the repetition rate with a synthesized microwave frequency at 980 MHz. Recall that simply shining comb light onto a high bandwidth photodiode generates the repetition rate (and its harmonics) as a beat of the comb with itself. Higher harmonics of the repetition rate magnify the detected frequency jitter, allowing the fundamental frequency to be determined within a tighter range, potentially below the counter resolution. The frequency counter records this *down-mixed* repetition rate at *approximately* 20 MHz (since $f_{\rm RR} \approx 250$ MHz), and this should not be confused with the beat of the master laser and the comb.

As expected,[27] $f_{\rm CEO}$ averages as white phase noise ($\sigma_y(\tau) \propto \tau^{-1}$ where τ is the averaging time), reaching a fractional instability of 10^{-13} in 10^4 s. The optically stabilized repetition rate, $f_{\rm RR}$, averages slightly slower and reaches a floor of $\sim 2 \times 10^{-13}$ before rising again. This long-term behavior is almost certainly due to a combination of residual cavity drift not compensated by the ramped AOM between the master laser and the reference cavity (see the previous subsection) and the drift of the rubidium standard (through the frequency counter). Both the "ultralow noise" frequency comb and the frequency counter contribute negligible noise at this level.

As an aside, we note that techniques to cancel out comb noise exist. In fact, a variation was implemented in Ref. [17] to suppress the noise of an older comb model we had at that time. These *transfer oscillator* schemes [18–20] typically involve real-time cancellation of the carrier-envelope offset and the repetition rate from the laser-comb beats. In the simplest realization, the carrier-envelope offset is mixed out from the master laser-comb beat (e.g., $f_{\rm b,1} + s f_{\rm CEO}$), and the resulting frequency (e.g., $f_{\rm laser,\,1} - N_1 f_{\rm RR}$) is plugged into the reference port of a DDS with high bit count. The tuning word for the DDS is set to synthesize the frequency $(N_2/N_1) \times (f_{\rm laser,\,1} - N_1 f_{\rm RR}) = (N_2/N_1) f_{\rm laser,1} - N_2 f_{\rm RR}$. Next, the carrier-envelope offset for the slave laser-comb beat with the same sign is similarly subtracted, and the resulting frequency $(f_{\rm laser,\,2} - N_2 f_{\rm RR})$ is phase-locked to the DDS synthesized frequency by actuating on the slave laser. Thus, a direct relationship between the master and slave laser is established, with the comb frequencies fully subtracted out; i.e., $f_{\rm laser,\,2} = (N_2/N_1) f_{\rm laser,1}$. In practice, this is limited by the bit count of the DDS in approximating the scaling factor N_2/N_1, and the mixing process may introduce additional phase noise.

[27] Because both $f_{\rm CEO}$ and the counter are phase referenced to the same rubidium standard.

3.2 Toy Model of Light–Molecule Interactions

3.2.1 Rotating Wave Hamiltonian

The archetypal level structure in this thesis involves at most two non-degenerate $J = 0$ ground states and three excited sub-levels belonging to a $J' = 1$ state. We turn to the optical Bloch equations (OBE) to model light-molecule interactions.

The Hamiltonian governing the unitary dynamics in the electric dipole ($E1$) and rotating-wave approximation can be written as

$$\mathbf{H}_{\text{rwa}} = \begin{pmatrix} \delta_1 & 0 & \frac{\Omega_1}{2}\epsilon_{(1),-1} & \frac{\Omega_1}{2}\epsilon_{(1),0} & \frac{\Omega_1}{2}\epsilon_{(1),+1} & 0 \\ 0 & \delta_2 & \frac{\Omega_2}{2}\epsilon_{(2),-1} & \frac{\Omega_2}{2}\epsilon_{(2),0} & \frac{\Omega_2}{2}\epsilon_{(2),+1} & 0 \\ \left(\frac{\Omega_1}{2}\epsilon_{(1),-1}\right)^* & \left(\frac{\Omega_2}{2}\epsilon_{(2),-1}\right)^* & -Z_{\text{split}} & 0 & 0 & 0 \\ \left(\frac{\Omega_1}{2}\epsilon_{(1),0}\right)^* & \left(\frac{\Omega_2}{2}\epsilon_{(2),0}\right)^* & 0 & 0 & 0 & 0 \\ \left(\frac{\Omega_1}{2}\epsilon_{(1),+1}\right)^* & \left(\frac{\Omega_2}{2}\epsilon_{(2),+1}\right)^* & 0 & 0 & +Z_{\text{split}} & 0 \\ 0 & 0 & 0 & 0 & 0 & 0 \end{pmatrix}. \tag{3.6}$$

As illustrated in Fig. 3.7, $\{|1\rangle, |2\rangle\}$ represent the ground states with angular momentum $J = 0, M = 0$, while $\{|3\rangle, |4\rangle, |5\rangle\}$ represents the excited Zeeman sub-levels with angular momentum $J' = 1, M' = -1, 0, +1$, separated by the Zeeman splitting Z_{split} in angular frequency units. Note that M and M' are the projection of the respective total angular momentum onto the lab-frame quantization

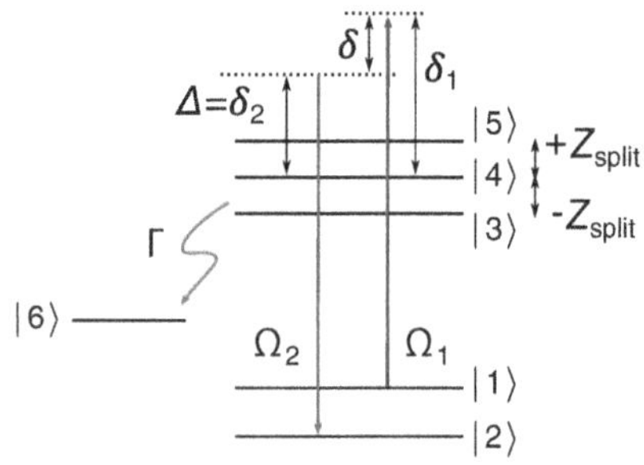

Fig. 3.7 Level diagram and relevant couplings for the optical Bloch equation simulation

axis.[28] We further augment the state space with $|6\rangle$, whose sole purpose is to collect population decay from $|k\rangle$, where $k \in \{3, 4, 5\}$. This is typical of molecular rovibronic transitions having non-diagonal Frank-Condon factors.

We introduce the electric fields of the upleg (pump) and downleg (anti-Stokes) lasers

$$\mathcal{E}_j = \frac{\mathcal{E}_{0,j}}{2}\left(\boldsymbol{\epsilon}_{(j)}\, e^{-i\omega_j t} + \boldsymbol{\epsilon}^*{}_{(j)}\, e^{+i\omega_j t}\right), \tag{3.7}$$

where $j \in \{1, 2\}$, ω_j is the optical angular frequency, $\mathcal{E}_{0,j}$ is the electric field amplitude, and $\delta_j \equiv \omega_j - \omega_{0,j}$ is the angular frequency detuning relative to the $M' = 0$ sub-level with transition frequency $\omega_{0,j}$ from the corresponding ground state. We pick the convention that the lasers propagate along the positive $\hat{\mathbf{Y}}$-direction and write the polarization vectors as

$$\boldsymbol{\epsilon}_{(j)} = \hat{\mathbf{Z}} \cos\theta_j + \hat{\mathbf{X}} e^{i\phi_j} \sin\theta_j, \tag{3.8}$$

where θ_j is the polar angle relative to $\hat{\mathbf{Z}}$ and the phase ϕ_j parameterizes the degree of circular polarization.

As in quantum optics, the term $\boldsymbol{\epsilon}\, e^{-i\omega t}$ corresponds to the annihilation operator, while its complex conjugate $\boldsymbol{\epsilon}^*\, e^{+i\omega t}$ corresponds to the creation operator. This distinction is the reason why there is a difference in the polarization dependence of the effective two-photon Rabi frequency for a Raman transition (Λ-system) versus a consecutive two-photon transition (Ξ-system); see e.g., Ref. [22]. Since the transition from $|j\rangle$ to $|k\rangle$ involves the annihilation of a photon, the angular Rabi frequency[29] is

$$\Omega_{j,k} = \frac{\mathcal{E}_{0,j}}{\hbar}\, \langle k|\mathbf{d} \cdot \boldsymbol{\epsilon}_{(j)}|j\rangle, \tag{3.9}$$

$$= \frac{\mathcal{E}_{0,j}}{\hbar} \sum_p (-1)^p \epsilon_{(j),-p} \langle k|d_p^{(1)}|j\rangle, \tag{3.10}$$

[28] The relative magnitudes of the trap depth and the induced Zeeman splitting determine the quantization axis. For example, the quantization axis of $X(v, J)$ states are almost always defined by the lattice polarization because these states have a negligible magnetic moment. On the other hand, the excited states of $(1)0_u^+$ and $(1)1_u$ are magnetic, and it is more appropriate to take their quantization axis to be that of the applied magnetic field. We say there is mixed quantization when the lattice polarization is not parallel (e.g. perpendicular) to the applied magnetic field. This situation can lead to apparent violations of $E1$ selection rules [21].

[29] In colloquial speech, the angular sense of this rate quantity (because of the division of [Energy] by the reduced Planck constant $\hbar$) is understood under the context in which it occurs, and is just referred to as the "Rabi frequency".

$$= \frac{\mathcal{E}_{0,j}}{\hbar} \langle k||d^{(1)}||j\rangle \sum_p (-1)^{p+1-M'_k} \epsilon_{(j),-p} \begin{pmatrix} 1 & 1 & 0 \\ -M'_k & p & 0 \end{pmatrix}, \tag{3.11}$$

$$= \frac{\mathcal{E}_{0,j}}{\hbar} \frac{\langle k||d^{(1)}||j\rangle}{\sqrt{3}} \epsilon_{(j),M'_k}. \tag{3.12}$$

In the second step, we express the dot product in terms of the spherical tensors $d_p^{(1)}$ and $\epsilon_{(j),-p}$, with $p \in \{0, \pm 1\}$. In the third step, we used the **Wigner-Eckart theorem** for a transition mediated by the multipole operator (**Q**) of rank L with component K between two states labeled generically as α and β,

$$\langle \beta; J', M'|Q_K^{(L,\lambda)}|\alpha; J, M\rangle = (-1)^{J'-M'} \begin{pmatrix} J' & L & J \\ -M' & K & M \end{pmatrix} \langle \beta(J')||Q^{(L,\lambda)}||\alpha(J)\rangle, \tag{3.13}$$

where the array with round brackets is the Wigner 3-j symbol. The double-barred $\langle \beta(J')||Q^{(L,\lambda)}||\alpha(J)\rangle$ is called the reduced matrix element, and it is simply a number (so it has no components). Here, λ serves as a clarification symbol and is not a quantum number. $\lambda = 1(0)$ for electric (magnetic) multipoles; i.e., the electric dipole operator $d_p^{(1)}$ corresponds to $Q_p^{(1,1)}$. The effect of centrifugal distortion on the spatial wavefunctions is implied by the $\alpha(J)$, $\beta(J')$ labels. The spherical polarization components of the laser addressing the state $|j\rangle$ can be written as

$$\epsilon_{(j),0} = \cos\theta_j, \ \epsilon_{(j),\pm 1} = \mp \frac{1}{\sqrt{2}} e^{i\phi_j} \sin\theta_j, \tag{3.14}$$

where the angles are defined the same as in Eq. (3.8). Note that $(\epsilon^*)_p = (-1)^p(\epsilon_{-p})^*$ is not equal to $(\epsilon_p)^*$, and that $\boldsymbol{\epsilon}^* \cdot \boldsymbol{\epsilon} = \sum_p (-1)^p (\epsilon^*)_p \epsilon_{-p} = 1$. In the final step, we make use of the fact that the 3-j symbol is zero except when $p = M'_k$, and that the square of the 3-j symbol is independent of M' for the special case $J = 0$ and $J' = 1$,

$$\begin{pmatrix} 1 & 1 & 0 \\ -M' & M' & 0 \end{pmatrix}^2 = \frac{1}{3}. \tag{3.15}$$

Equation (3.15) is the reason for the intuitive form of the off-diagonal coupling terms ($\Omega_{j,k}/2$) in our Hamiltonian [Eq. (3.6)]. That is, defining

$$\Omega_j \equiv \frac{\mathcal{E}_{0,j}}{\hbar} \frac{\langle k||d^{(1)}||j\rangle}{\sqrt{3}}, \tag{3.16}$$

we see that the angular Rabi frequencies $\Omega_{j,k} = \Omega_j \epsilon_{(j),M'_k}$ (note that $\epsilon_{-p=\pm 1} = -\epsilon_{p=\pm 1}$).

Transition rates, Autler-Townes splittings, and light shifts grant experimental access to the quantity $\sum_k |\Omega_{j,k}|^2$ when the excited Zeeman sub-levels are unresolved, or when the laser is far detuned that the sub-levels are effectively degenerate. This presents an advantage because $\sum_k |\Omega_{j,k}|^2 = |\Omega_j|^2$ is polarization independent for $J = 0 \rightarrow J' = 1$.

3.2.2 *Lindblad Master Equation*

The Lindblad master equation for our system is

$$\frac{\mathrm{d}}{\mathrm{d}t}\rho = -i[\mathbf{H}_{\mathrm{rwa}}, \rho] + \mathcal{D}_{1,2}(\rho) + \sum_{k=3,4,5} \mathcal{L}_k(\rho), \tag{3.17}$$

where [,] is the commutator.

To put in the relaxation from $|k\rangle$ to the auxiliary state $|6\rangle$, we compute

$$\mathcal{L}_k(\rho) = -\frac{1}{2}\{\mathbf{G}_\mathbf{k}^\dagger \mathbf{G}_\mathbf{k}, \rho\} + \mathbf{G}_\mathbf{k}\rho\mathbf{G}_\mathbf{k}^\dagger, \tag{3.18}$$

where {,} is the anti-commutator, $\mathbf{G}_\mathbf{k}^\dagger \equiv \sqrt{\Gamma}|k\rangle\langle 6|$ are the so-called *jump operators*,[30] Γ is the excited state linewidth (full width at half maximum) in angular frequency units,[31] and ρ is the 6×6 density matrix for the entire state space. Basically, $\mathcal{L}_k$ generates decay terms proportional to $-\Gamma$ in the diagonals of $|k\rangle$ and $-\Gamma/2$ in the off-diagonals between $|k\rangle$ and $|6\rangle$.

In some cases, it is useful to include a phenomenological decoherence rate Γ_{eff} which can be interpreted as the relative linewidth between the upleg and downleg. This is done by computing

$$\mathcal{D}_{1,2}(\rho) = -\frac{\Gamma_{\mathrm{eff}}}{2}\left(\mathbf{P}_1\rho\mathbf{P}_2 + \mathbf{P}_2\rho\mathbf{P}_1\right), \tag{3.19}$$

where $\mathbf{P}_\mathbf{j} \equiv |j\rangle\langle j|$ are projection operators onto the corresponding diagonal element. $\mathcal{D}_{1,2}$ will generate decay terms proportional to $-\Gamma_{\mathrm{eff}}/2$ in the off-diagonals between $|1\rangle$ and $|2\rangle$.

The best way to approach this is to numerically solve the time evolution of Eq. (3.17) for the input parameters Ω_j, δ_j, θ_j, ϕ_j, Z_{split}, Γ, and Γ_{eff}, with the initial

[30] Note that there is a typo in the expression for the jump operator in Ref. [23]; there should be a Hermitian conjugate.

[31] The lifetime is $1/\Gamma$. Whenever in doubt about factors of 2π concerning frequencies or lifetimes, revert back to the elementary example of a damped driven harmonic oscillator in classical mechanics.

condition $\rho_{11}(t=0)=1$ and zero for all other entries. Nevertheless, it is instructive to examine several limiting cases to build our intuition.

3.3 One-Photon Spectroscopy, the Scattering Lineshape

We first examine the simple case of one-photon spectroscopy. Here, $\Omega_2 = 0$, $\Gamma_{\text{eff}} = 0$ and population is initially in state $|1\rangle$. Writing out the master equation [Eq. (3.17)], we arrive at the following set of coupled differential equations,

$$\frac{\mathrm{d}}{\mathrm{d}t}\rho_{11} = -\sum_k \mathrm{Im}\left(\Omega_{1,k}^* \rho_{1k}\right), \tag{3.20}$$

$$\frac{\mathrm{d}}{\mathrm{d}t}\rho_{1k} = \left(-\frac{\Gamma}{2} - i(\delta_1 - Z_{\text{split},k})\right)\rho_{1k} - \frac{i}{2}\Omega_{1,k}(\rho_{kk} - \rho_{11}), \tag{3.21}$$

$$\frac{\mathrm{d}}{\mathrm{d}t}\rho_{kk} = +\mathrm{Im}\left(\Omega_{1,k}^* \rho_{1k}\right) - \Gamma\rho_{kk}, \tag{3.22}$$

where $Z_{\text{split},k} = -Z_{\text{split}}, 0, +Z_{\text{split}}$ for $k = 3, 4, 5$ respectively, and Im() means the "imaginary part".

Since Γ for deeply bound 0_u^+ states is typically larger than Ω, hardly any population accumulates in $|k\rangle$, and we may approximate $\rho_{kk} \approx 0$. Open systems do not have a well-defined "steady state"; as $t \to \infty$, all population gets optically pumped into $|6\rangle$. Nevertheless, we are certainly allowed to analyze the situation *approaching* quasi-steady-state conditions, where we set $\mathrm{d}\rho_{1k}/\mathrm{d}t$ to zero and solve for ρ_{1k} using Eq. (3.21). Substituting the result into Eq. (3.20) and taking the special case where $|Z_{\text{split}}| \ll \Gamma$ (unresolved Zeeman sub-levels) we get,

$$\frac{\mathrm{d}}{\mathrm{d}t}\rho_{11} = -\frac{\Gamma\,|\Omega_1|^2}{\Gamma^2 + 4\delta_1^2}\,\rho_{11}. \tag{3.23}$$

Therefore, the one-photon scattering rate (summed over all sub-levels) due to the upleg laser is

$$R_{\text{sc}} = \frac{\Gamma\,|\Omega_1|^2}{\Gamma^2 + 4\delta_1^2}. \tag{3.24}$$

The population in the initial state ($N \equiv \rho_{11}$) thus decays as

$$N(\delta_1, t) = N_0\,\exp\left[-R_{\text{sc}}\,t\right] = N_0\,\exp\left[-t\,\frac{\Gamma\,|\Omega_1|^2}{\Gamma^2 + 4\delta_1^2}\right]. \tag{3.25}$$

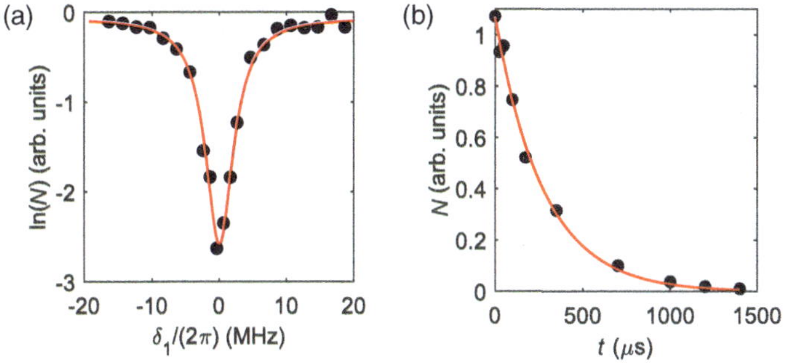

Fig. 3.8 Black points are single instances of the experiment. Solid red lines are simultaneous fits according to Eq. (3.25). (**a**) One-photon lineshape for a bound-to-bound transition, shown for $X(62, 0) \rightarrow (1)0_u^+(12, 1)$ at 789 nm. Here, we take the natural logarithm of the spectroscopy signal (proportional to the number of molecules). The fit function has the form $y = -t\Gamma|\Omega_1|^2/(\Gamma^2 + 4(x - x_c)^2) + \text{const.}$, with t fixed. (**b**) One-photon scattering near resonance ($\delta_1 \approx 0$ as close as possible). The number of molecules decays at the rate of the driven excitation. The fit function has the form $y = \text{const.} \times \exp[-t\Gamma|\Omega_1|^2/(\Gamma^2 + 4(x - x_c)^2)]$, with x fixed. The fitting is performed *simultaneously* in both plots; i.e., the bound-to-bound laser power is kept the same between both sets of data, and the fits share the global parameters Γ, $|\Omega_1|^2$, and x_c (the fit line center). Note that $(x - x_c) = (\omega_1 - \omega_{0,1}) \equiv \delta_1$

For cases where the states are non-degenerate ($|Z_{\text{split}}| > \Gamma$), we simply replace $\Omega_1 \rightarrow \Omega_{1,k}$ and $\delta_1 \rightarrow (\delta_1 - Z_{\text{split},k})$ and sum over k to compute R_{sc} in Eq. (3.24). As shown in Fig. 3.8, simultaneous fits to Eq. (3.25) of a peak scan (t kept constant, δ_1 varied) and a decay curve (t varied, $\delta_1 \approx 0$ kept constant) allow the accurate extraction of $|\Omega_1|^2$, Γ, and the line center (where $\delta_1 = 0$).

3.4 Two-Photon Raman Spectroscopy

3.4.1 Far Detuned Limit, Effective Rabi Frequency, and Rabi Oscillations

We now focus on two-photon spectroscopy and reinstate Ω_2 and Γ_{eff}. The master equation [Eq. (3.17)] gives the following set of coupled differential equations,

$$\frac{\mathrm{d}}{\mathrm{d}t}\rho_{11} = -\sum_k \operatorname{Im}\left(\Omega_{1,k}^* \rho_{1k}\right), \tag{3.26}$$

$$\frac{\mathrm{d}}{\mathrm{d}t}\rho_{1k} = \left(-\frac{\Gamma}{2} - i(\delta_1 - Z_{\text{split},k})\right)\rho_{1k} - i\frac{\Omega_{1,k}}{2}\left(\rho_{kk} - \rho_{11}\right) + i\frac{\Omega_{2,k}}{2}\rho_{12}, \tag{3.27}$$

$$\frac{\mathrm{d}}{\mathrm{d}t}\rho_{12} = \left(-\frac{\Gamma_{\mathrm{eff}}}{2} - i\delta\right)\rho_{12} + i\sum_k \frac{\Omega_{2,k}^*}{2}\rho_{1k} - i\sum_k \frac{\Omega_{1,k}}{2}\rho_{k2}, \tag{3.28}$$

$$\frac{\mathrm{d}}{\mathrm{d}t}\rho_{2k} = \left(-\frac{\Gamma}{2} - i(\delta_2 - Z_{\mathrm{split},k})\right)\rho_{2k} - i\frac{\Omega_{2,k}}{2}(\rho_{kk} - \rho_{22}) + i\frac{\Omega_{1,k}}{2}\rho_{21}, \tag{3.29}$$

$$\frac{\mathrm{d}}{\mathrm{d}t}\rho_{22} = -\sum_k \mathrm{Im}\left(\Omega_{2,k}^*\rho_{2k}\right), \tag{3.30}$$

$$\frac{\mathrm{d}}{\mathrm{d}t}\rho_{kk} = +\mathrm{Im}\left(\Omega_{1,k}^*\rho_{1k}\right) + \mathrm{Im}\left(\Omega_{2,k}^*\rho_{2k}\right) - \Gamma\rho_{kk}. \tag{3.31}$$

We define the one-photon (common) detuning $\Delta \equiv \delta_2$, and the two-photon (Raman) detuning $\delta \equiv \delta_1 - \delta_2$. The reader is reminded that these are angular frequencies. Again, we expect $\rho_{kk} \approx 0$ as the population will either be transferred to $|2\rangle$ or optically pumped to $|6\rangle$. In the far detuned limit where $\delta_1 \approx \delta_2 \equiv \Delta \gg |\Omega_1|, |\Omega_2|, \Gamma, |Z_{\mathrm{split}}|$, we expect the rate of change of the coherences $\mathrm{d}\rho_{1k}/\mathrm{d}t \approx 0$ and $\mathrm{d}\rho_{2k}/\mathrm{d}t \approx 0$. This simplifies the situation, allowing us to solve for ρ_{1k} and ρ_{k2} using Eqs. (3.27) and (3.29),

$$\rho_{1k} \approx \frac{1}{2\Delta}\left(\Omega_{1,k}\rho_{11} + \Omega_{2,k}\rho_{12}\right), \tag{3.32}$$

$$\rho_{k2} = \rho_{2k}^* \approx \frac{1}{2\Delta}\left(\Omega_{2,k}^*\rho_{22} + \Omega_{1,k}^*\rho_{12}\right). \tag{3.33}$$

Substituting Eqs. (3.32) and (3.33) back into Eq. (3.28), we obtain

$$\frac{\mathrm{d}}{\mathrm{d}t}\rho_{12} \approx \left[-\frac{\Gamma_{\mathrm{eff}}}{2} - i\left(\delta + \frac{|\Omega_1|^2}{4\Delta} - \frac{|\Omega_2|^2}{4\Delta}\right)\right]\rho_{12} - \frac{i}{2}\left[\sum_k \frac{\Omega_{2,k}^*\Omega_{1,k}}{2\Delta}\right](\rho_{22} - \rho_{11}). \tag{3.34}$$

Comparing terms in Eq. (3.34) with the analogous Eq. (3.21), we identify the effective angular Rabi frequency between the ground states $|1\rangle$ and $|2\rangle$ in our toy model,

$$\Omega_{\mathrm{eff}} \equiv \sum_k \frac{\Omega_{2,k}^*\Omega_{1,k}}{2\Delta} = \sum_k \frac{\langle 2|\mathbf{d}\cdot\boldsymbol{\epsilon}^*{}_2|k\rangle\langle k|\mathbf{d}\cdot\boldsymbol{\epsilon}_1|1\rangle}{2\Delta}. \tag{3.35}$$

The appearance of $\boldsymbol{\epsilon}^*$ for the second Raman step is because the downleg creates a photon (see also the footnote on photon annihilation for Eq. (3.9)). From here, it is easy to see that for the realistic case of many v', J', M' excited levels, the **effective angular Rabi frequency for a Raman transition** is

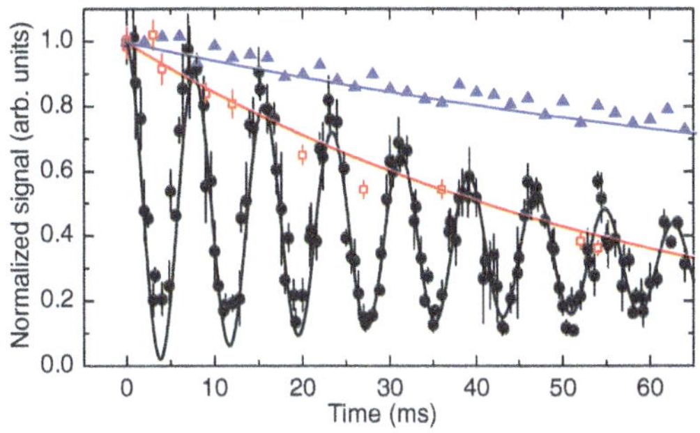

Fig. 3.9 Two-photon Rabi oscillations between $X(62, 0)$ and $X(4, 0)$ (black circles) in a magic wavelength optical lattice, averaged over 14 consecutive experimental traces. Also shown are the normalized population decay of $X(4, 0)$ (red squares) and $X(62, 0)$ (blue triangles) under the same experimental conditions. Black line: fit in the form $y = A\exp(-x/T_1)[1 + \exp(-x/T_2^{\mathrm{Rabi}})\cos(\omega x - \phi)]$, with $A, T_1, T_2^{\mathrm{Rabi}}, \omega, \phi$ as free parameters. Red and blue lines are fitted solutions of the rate equation $\mathrm{d}N/\mathrm{d}t = -k_\gamma N^\gamma$ with $\gamma = 1$ and 2 respectively. N is the molecular number, and k_γ is a free parameter

$$\Omega_{\mathrm{eff}} = \sum_f \frac{\langle 2|\mathbf{d}\cdot\boldsymbol{\epsilon}^*{}_2|f\rangle\langle f|\mathbf{d}\cdot\boldsymbol{\epsilon}_1|1\rangle}{2\Delta_f}, \tag{3.36}$$

where $|f\rangle$ is an excited intermediate state, and Δ_f is the detuning of the Raman probes relative to $|f\rangle$, addressed from $|1\rangle$, $|2\rangle$. Indeed, on Raman resonance ($\delta = 0$, ignoring the light shifts), the ground states $|1\rangle$ and $|2\rangle$ undergo Rabi oscillations and the population in the initial state as a function of time is

$$N(\delta = 0, t) \simeq \frac{N_0}{2}\exp\left(-t/T_1\right)\left[1 + \exp\left(-t/T_2^{\mathrm{Rabi}}\right)\cos\left(|\Omega_{\mathrm{eff}}|\, t\right)\right], \tag{3.37}$$

where we have inserted an overall empirical decay factor to account for finite state lifetimes $\sim T_1$, and define the Rabi "coherence time" $T_2^{\mathrm{Rabi}} \simeq 1/\Gamma_{\mathrm{eff}}$. Figure 3.9 depicts two-photon Rabi oscillations between $X(62, 0)$ and $X(4, 0)$. Here, $T_1 = 127(8)$ ms, and $T_2^{\mathrm{Rabi}} = 77(6)$ ms (or equivalently, 19(1) cycles). The oscillations are predominately damped by the loss of $X(4, 0)$ molecules, which has a $1/e$ lifetime of 60(2) ms due to lattice-induced one-body losses under the conditions of that experiment (see also Sect. 5.3.3). T_1 is roughly twice as long as the $1/e$ lifetime of $X(4, 0)$ because the molecules only spend, on average, half the time in this state as they undergo Rabi oscillations.

Another important takeaway from Eq. (3.34) is its prediction that the Raman resonance will be light shifted[32] by an angular frequency $|\Omega_1|^2/(4\Delta)$ due to the upleg, and $-|\Omega_2|^2/(4\Delta)$ due to the downleg. In this toy model with just one $J' = 1$

[32] See also Eq. (3.46).

excited state, the total Raman probe light shift is nulled when the Rabi frequencies are balanced (i.e., $|\Omega_1| = |\Omega_2|$).

3.4.2 Weak Probe Limit, Dark Resonance, and Autler-Townes Doublet

Close to one-photon resonance ($\Delta \approx 0$) and in the weak probe limit ($|\Omega_1| \ll |\Omega_2|$), we may approximate $\rho_{11} \approx 1$, $\rho_{22} \approx 0$, and $\rho_{kk} \approx 0$ and we can derive a general analytical expression for the excitation lineshape when both Raman lasers are applied. Again we examine the simple case where $|Z_{\text{split}}| \ll \Gamma$ (unresolved Zeeman sub-levels) so that effectively $Z_{\text{split}} \approx 0$. Discarding negligible terms $O(\Omega_1/\Omega_2)$,

$$\frac{\mathrm{d}}{\mathrm{d}t}\rho_{11} \approx -\sum_k \mathrm{Im}\left(\Omega_{1,k}^*\rho_{1k}\right), \tag{3.38}$$

$$\frac{\mathrm{d}}{\mathrm{d}t}\rho_{1k} \approx \left(-\frac{\Gamma}{2} - i\delta_1\right)\rho_{1k} + i\frac{\Omega_{1,k}}{2} + i\frac{\Omega_{2,k}}{2}\rho_{12}, \tag{3.39}$$

$$\frac{\mathrm{d}}{\mathrm{d}t}\rho_{12} \approx \left(-\frac{\Gamma_{\text{eff}}}{2} - i\delta\right)\rho_{12} + i\sum_k \frac{\Omega_{2,k}^*}{2}\rho_{1k}, \tag{3.40}$$

Since here $\rho_{11} \approx 1$, the first equation immediately implies that the rate of excitation out of $|1\rangle$ is $R_{\text{sc}} = \sum_k \mathrm{Im}\left(\Omega_{1,k}^*\rho_{1k}\right)$. Again, to find the excitation rate approaching steady-state conditions, we set $\mathrm{d}\rho_{12}/dt \approx \mathrm{d}\rho_{1k}/dt \approx 0$. We first solve for ρ_{12} using Eq. (3.40) and substitute the result into Eq. (3.39). Then, multiplying through by $\Omega_{2,k}^*$ and taking the sum over k allows one to solve for $\sum_k \Omega_{2,k}^*\rho_{1k}$, which we substitute back into Eq. (3.39) via the expression for ρ_{12}. Finally, we solve for ρ_{1k} in terms of the Rabi frequencies, detunings, and linewidths.

Using $R_{\text{sc}} = \sum_k \mathrm{Im}\left(\Omega_{1,k}^*\rho_{1k}\right)$, we find

$$\begin{aligned} R_{\text{sc}} = & \frac{\Gamma}{\Gamma^2 + 4\delta_1^2} \\ & \times \left[|\Omega_1|^2 - \left(\sum_k \Omega_{1,k}^*\Omega_{2,k}\right)\left(\sum_k \Omega_{1,k}\Omega_{2,k}^*\right)\right. \\ & \left. \times \frac{|\Omega_2|^2 - 8\delta\delta_1 + \Gamma_{\text{eff}}\Gamma(1 - 4\delta_1^2/\Gamma^2)}{||\Omega_2|^2 + (\Gamma + 2i\delta_1)(\Gamma_{\text{eff}} + 2i\delta)|^2}\right]. \end{aligned} \tag{3.41}$$

Note the absolute square in the denominator of the fraction can be expanded to

$$(|\Omega_2|^2 + \Gamma\Gamma_{\text{eff}} - 4\delta_1\delta)^2 + 4(\Gamma\delta + \Gamma_{\text{eff}}\delta_1)^2.$$

Equation (3.41) has been verified against numerical solutions of the master equation [Eq. (3.17)]. It is also immediately clear from the second term of Eq. (3.41), or more fundamentally from Eq. (3.36), that the Raman, two-photon effect vanishes for our case ($J = 0 \leftrightarrow J' = 1$) when the Raman lasers have orthogonal polarizations.

Taking the exponential just as before, the general lineshape in the weak probe limit is

$$N(\delta, \delta_1, t) = N_0 \exp\left[-R_{\text{sc}}t\right]. \tag{3.42}$$

For the ideal case where the Raman lasers have exactly the same polarization (i.e. $\theta_1 = \theta_2$, $\phi_1 = \phi_2$), we have $\left(\sum_k \Omega^*_{1,k}\Omega_{2,k}\right)\left(\sum_k \Omega_{1,k}\Omega^*_{2,k}\right) = |\Omega_1|^2|\Omega_2|^2$ since both ground states have $J = 0$. Thus Eqs. (3.41) and (3.42) further simplify to

$$N(\delta, \delta_1, t) = N_0 \exp\left[-t\,\frac{\Gamma|\Omega_1|^2}{\Gamma^2 + 4\delta_1^2}\left(1 - |\Omega_2|^2\,\frac{|\Omega_2|^2 - 8\delta\delta_1 + \Gamma_{\text{eff}}\Gamma(1 - 4\delta_1^2/\Gamma^2)}{\left||\Omega_2|^2 + (\Gamma + 2i\delta_1)(\Gamma_{\text{eff}} + 2i\delta)\right|^2}\right)\right], \tag{3.43}$$

where the single-photon and two-photon effects are separated into distinct terms. If we set $|\Omega_2| = 0$, we immediately recover Eq. (3.25). Equation (3.43) is also valid in the case where the excited state structure is very well separated such that one effectively addresses a three-level Λ-system.

As shown in Fig. 3.10, Eq. (3.43) is an excellent fit to the experimental data. Here, the upleg addresses $X(62, 0) \to (1)0_u^+(11, 1)$ and the downleg addresses $X(0, 0) \to (1)0_u^+(11, 1)$. For $|\Omega_2| \ll \Gamma$ and pulse durations within the Raman coherence time ($t \lesssim 1/\Gamma_{\text{eff}}$), as in Fig. 3.10a, we observe the appearance of a narrow peak embedded within the broader one-photon depletion when we scan the upleg across the excited state.[33] This phenomenon is called **electromagnetically induced transparency (EIT)**. The narrow peak[34] is called a **dark resonance** which occurs when the two-photon condition is met ($\delta \approx 0$).

For increasingly large downleg laser powers ($|\Omega_2| \gg \Gamma$), as in Fig. 3.10b, c, d, and e, the EIT lineshape progressively becomes an **Autler-Townes doublet**. The

[33] As mentioned, the sharpness and "height" of the EIT peak depends on the level of light-molecule coherence. As such, the lattice light will have to be shuttered off momentarily for a non-magic trap during the Raman pulse to eliminate differential lattice-induced light shifts. Typical pulse durations to achieve results like Fig. 3.10a are $\lesssim 100\,\mu$s. Additionally, the relative polarizations of the up and down leg should be made as parallel as possible.

[34] Often the dark resonance may appear asymmetric due to lattice or probe light shifts, or if the downleg is slightly detuned from the excited state. In the latter case, the lineshape resembles a Fano resonance.

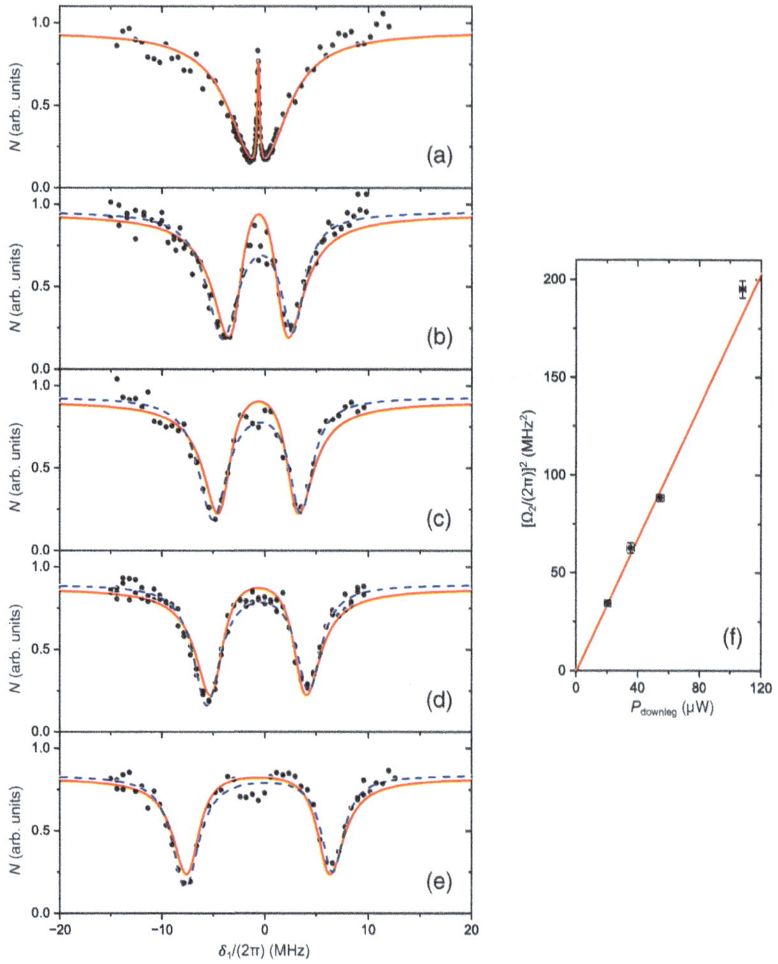

Fig. 3.10 Progression from EIT (free space) to the Autler-Townes doublet. Black points, single shot of the experiment. Red lines, fits to Eq. (3.43). The EIT fit in (**a**) only has two free parameters; δ_2, Γ_{eff}. This is because t is set by the experimental sequence, and $|\Omega_1|$, Γ and the line center (where $\delta_1 = 0$) are determined from one-photon spectroscopy. $|\Omega_2|$ in (**a**) is extrapolated from (**f**), a linear fit of $|\Omega_2|^2$ versus the downleg power. The respective $|\Omega_2|$ for the Autler-Townes doublets (**b**), (**c**), (**d**), (**e**) may be extracted by fitting either to Eq. (3.43) as an additional free parameter, or to the sum of two Lorentzians (blue dashed lines), which is a valid approximation when $|\Omega_2| \gg \Gamma$

Autler-Townes effect can be easily understood in the dressed state picture. The downleg laser dresses $|2\rangle$, forming a pair of dressed states with $|4\rangle$ (ignoring $|3\rangle$ and $|5\rangle$ for simplicity). As a generic feature of quantum mechanics, the coupling of $|2\rangle$ and $|4\rangle$ by the downleg repels the dressed states.

To understand this analytically, consider the effective Hamiltonian between $|2\rangle$ and $|4\rangle$,

$$H_2 = \begin{pmatrix} \Delta & \frac{\Omega_2}{2} \\ \frac{\Omega_2^*}{2} & 0 \end{pmatrix}.$$

Diagonalization of this Hamiltonian results in the eigenvalues of the dressed states

$$D_\pm = \frac{\Delta}{2} \pm \frac{\sqrt{|\Omega_2|^2 + \Delta^2}}{2}. \tag{3.44}$$

Depletion from $|1\rangle$ occurs when the weak upleg laser addresses the pair of dressed states, resulting in the appearance of a doublet in the spectrum. Let $D_+(D_-)$ be the line center of the right-side (left-side) peak relative to the bare resonance.[35] Hence, we see that

$$(D_+ - D_-)^2 = \Delta^2 + |\Omega_2|^2 \tag{3.45}$$

Keeping the downleg laser intensity constant, the square of the doublet separation $(D_+ - D_-)^2$ versus Δ is a parabola whose minimum is $|\Omega_2|^2$, as illustrated by Fig. 3.11a and b. Furthermore, the plot of $(D_+ - D_-)^2$ versus I_{downleg}, the intensity of the downleg (anti-Stokes) laser is a straight line with a vertical intercept equal to Δ^2 (see e.g., Fig. 3.10f but note $\Delta \equiv \delta_2 \approx 0$ for that dataset).

In either case, we have an all-frequency way[36] of determining $|\Omega_2|^2$ and the frequency splitting of the ground states,[37] especially for states that are not initially populated. Moreover, $(D_+ - D_-)^2$ is robust against various effects such as reference cavity drift, the overall trap-induced light shift, shot-to-shot signal fluctuations, and the fit lineshape for the doublet. By adjusting the experimental conditions (e.g., applied magnetic field strength) so that we measure the parabola in the regime where $|\Delta| > |Z_{\text{split}}|$, the extracted $|\Omega_2|^2$ is insensitive to laser polarization and are effectively between the $M = M' = 0$ sub-levels for $J = 0$ and $J' = 1$. Two-photon photoassociation (PA) starting from a sample of atoms is another example where Autler-Townes splittings were used to determine bound-to-bound molecular transition strengths [17, 24, 25]; e.g., Fig. 3.11a and b.

Lastly, we observe that taking the far detuned limit ($\Delta \gg |\Omega_2|$), Eq. (3.44) becomes

$$D_+ \approx \Delta + \frac{|\Omega_2|^2}{4\Delta}, \quad D_- \approx -\frac{|\Omega_2|^2}{4\Delta}. \tag{3.46}$$

[35] Bare resonance being the situation when $|\Omega_2| = 0$.

[36] Complementary to the method involving power broadening and transition rates described in Sect. 3.3.

[37] From the frequency difference of the up and downleg lasers at the molecules or atoms.

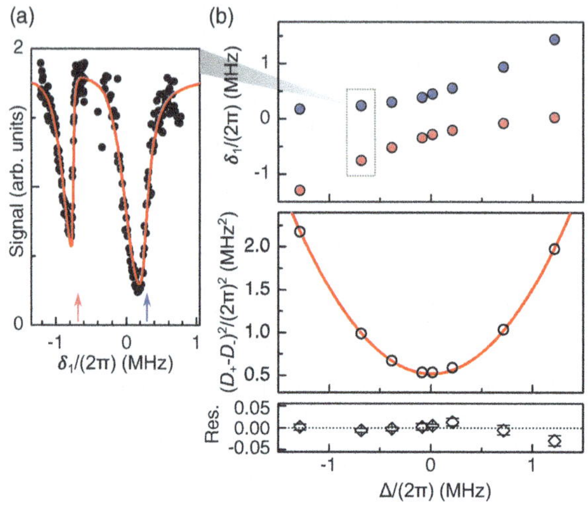

Fig. 3.11 Two-photon PA and Autler-Townes spectroscopy. (**a**) Autler-Townes doublet as the upleg is scanned across the free-to-bound PA transition ${}^1S_0 + {}^1S_0 \rightarrow 0_u^+(-4, 1)$, while keeping the downleg (anti-Stokes) at a fixed detuning, Δ, from the bound-to-bound transition $X(61, 0) \rightarrow 0_u^+(-4, 1)$. Here, we start with a sample of lattice-confined *atoms*. The fit function includes thermal broadening and skewing effects for PA in a 1D lattice; for details, see Ref. [21]. (**b**) Top: frequency splitting of the right- and left-side peaks (corresponding to the location of the color-coded arrows in the callout for a given Δ), illustrating the avoided crossing. Bottom: the square of the angular frequency splitting $(D_+ - D_-)^2$ plotted against Δ is a parabola with a minimum value of $|\Omega_2|^2$. Residuals have the same units as the plot

This result is general and implies that a laser, far detuned from a transition between two levels, induces an angular frequency shift equal to $+|\Omega_2|^2/(4\Delta)$ on the lower state, and $-|\Omega_2|^2/(4\Delta)$ on the upper state. Generalization to multiple excited levels naturally leads to the microscopic sum-over-states formula for polarizability (see Sect. 5.1.3).

3.5 Results and Analysis

3.5.1 $X^1\Sigma_g^+$

We perform dark-resonance spectroscopy to locate the bound states of $X^1\Sigma_g^+$ (or $X0_g^+$). The upleg is locked on resonance with $X(62, 0) \rightarrow (1)0_u^+(11, 1)$ and we adjust the laser power and pulse time such that we achieve nearly full depletion without bleaching the spectroscopic signal (photofragments of $X(62, 0)$). When the downleg is on resonance with a $X \rightarrow 0_u^+$ transition, $(1)0_u^+(11, 1)$ gets light shifted away, $X(62, 0)$ is no longer depleted, and we recover the spectroscopic signal.

Table 3.1 Precise sub-MHz vibrational splittings of the first nine irrotational ($J = 0$) levels of $X^1\Sigma_g^+$ with respect to $X(62, 0)$ in $^{88}Sr_2$. For $v = 0, 4, 6, 8$, we quote measurements from Raman clock spectroscopy in a magic wavelength lattice (see also Table 5.1). The rest are determined using dark resonance spectroscopy (line centers extrapolated to zero lattice intensity). Note that these vibrational splittings are *not* the binding energies. To obtain the latter, the binding energy of $X(62, 0)$ will have to be added to these splittings. Values are given in units of MHz

$X(v, 0)$	Vibrational splitting wrt. $X(62, 0)$
0	31,825,183.207 593(5)
1	30,640,159.793(25)
2	29,479,060.993(27)
3	28,341,865.710(26)
4	27,228,546.843(10)
5	26,139,071.677(26)
6	25,073,401.504(10)
7	24,031,492.503(26)
8	23,013,294.901(10)

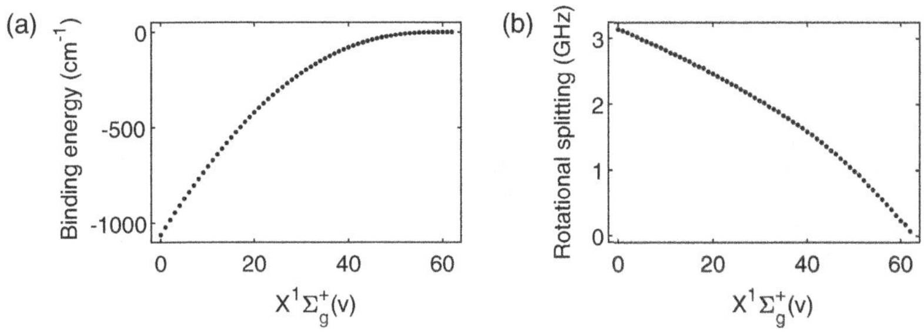

Fig. 3.12 Comprehensive laser spectroscopy of $X^1\Sigma_g^+$ in $^{88}Sr_2$. (**a**) The binding energies of all 63 vibrational states with $J = 0$. (**b**) The rotational splitting ($J = 2 - J = 0$) for each vibrational state. See also Table 3.4

For $v = 0$–8 and $J = 0$, we gradually lower the downleg power to further narrow the dark resonance, enabling the vibrational splittings relative to $X(62, 0)$ to be determined with sub-MHz accuracy. For these states, we corrected for the leading lattice light shift systematic and used a frequency comb to determine the corresponding difference frequencies of the Raman lasers. These **vibrational splittings** are reported in Table 3.1.

For spectroscopy of the remaining levels, we use a widely tunable Ti:sapphire laser as the downleg. To speed up the process of locating the states, here the downleg is rather intense ($\sim$500 W cm^{-2}), and we measure the vibrational splittings by reading off the Raman laser on a wavelength meter.

Ultimately, we successfully find **all** 63 vibrational states with $J = 0$ and 2 in $X^1\Sigma_g^+$ for the $^{88}Sr_2$ isotopologue. The **binding energies** relative to $^1S_0 + {}^1S_0$ and rotational splittings are shown in Fig. 3.12a and b, and listed in Table 3.4. To calculate the binding energies from the measured vibrational splittings, we take the binding energy of $X(62, 0)$ to be 136.6447(50) MHz from Ref. [26].

The short-range behavior of an isolated potential can be modeled with the simple Morse potential,

$$V_{\mathrm{M}}(R) = D_e\left[1 - e^{-\beta(R-R_e)}\right]^2 - D_e, \tag{3.47}$$

where D_e is the potential depth, $R_e = \frac{1}{2\pi}\sqrt{h/(2\mu_2 c B_e)}$ is the equilibrium bond length, $\beta \equiv 2\pi\sqrt{2\mu_2\omega_e x_e c/h}$, and μ_2 is the reduced mass of the dimer; i.e. $\mu_2 \equiv (1/m_{\mathrm{Sr}} + 1/m_{\mathrm{Sr}})^{-1} = m_{\mathrm{Sr}}/2$, where m_{Sr} is the mass of atomic ^{88}Sr.

In the presence of vibrational-rotational interaction, the energy levels can be characterized by the leading terms of the Dunham expansion (in units of cm^{-1}),

$$\begin{aligned} E(v, J) = &-D_e + \omega_e\left(v + \frac{1}{2}\right) - \omega_e x_e\left(v + \frac{1}{2}\right)^2 \\ &+\left[B_e - \alpha_e\left(v + \frac{1}{2}\right)\right]\left[J(J+1) - \Omega^2\right], \end{aligned} \tag{3.48}$$

where ω_e, x_e, B_e and α_e are the vibrational, anharmonicity, rotational and vibration-rotation coupling spectroscopic constants respectively. We expect this to be valid for deeply bound states of $X^1\Sigma_g^+$. The $J = 0, 2$ rotational splitting for $\Omega = 0$ is

$$E(v, 2) - E(v, 0) = 6\left[B_e - \alpha_e\left(v + \frac{1}{2}\right)\right]. \tag{3.49}$$

For a quick comparison of our spectroscopy with older studies involving Fourier transform spectra in a heatpipe [27, 28], we simultaneously fit the measured binding energies of the first nine levels ($v = 0$–8), plotted against $(v + 1/2)$, to Eqs. (3.48) and (3.49). The extracted spectroscopic constants are listed in Table 3.2. We find good agreement with the parameters reported in Refs. [27, 28] that are weighted across several isotopologues.

Table 3.2 Spectroscopic constants of $X^1\Sigma_g^+$ in $^{88}Sr_2$ estimated from fitting the measured binding energies of the first nine states ($v = 0$–8) to those of a simple vibrating-rotator

Spectroscopic constant	Value (cm^{-1})
D_e	−1081.6382(7)
ω_e	40.3186(6)
$\omega_e x_e$	−0.39746(7)
B_e	0.017579(6)
α_e	$-1.76(1)\times 10^{-4}$

3.5.2 $(1)0_u^+$

The quantum chemistry of the excited potentials is particularly rich, and previous work from our group studied the properties of weakly bound states near the intercombination line [7, 17, 21, 24, 25, 29]. In the current subsection and the next, we concentrate on the deeply bound states of $(1)0_u^+$ and $(1)1_u$.

Deeply bound $(1)0_u^+(J' = 1)$ states are accessible from weakly bound ground state molecules with laser light in the range ~700–850 nm. We perform one-photon spectroscopy to locate the states using a highly tunable Ti:sapphire laser. With a sample of $X(62, 0)$ molecules, we directly excite the first 37 vibrational levels of $(1)0_u^+$ with $J' = 1$. We further measure the binding energies of $v' = 9$–20 to sub-MHz accuracy with the frequency comb, scanning over each resonance with an acousto-optic modulator and finding the line centers with the method described in Sect. 3.3. All known states assigned to $(1)0_u^+$ with $J' = 1, 3$ are consolidated in Table 3.5.

There is no experimental hurdle to locating higher excited states beyond those found in this work. However, we have observed that additional, nearly degenerate states start to appear in the spectra around $-750\,\text{cm}^{-1}$. Some of these almost certainly belong to $(2)0_u^+$, as anecdotally, they are weaker, necessitating greater laser intensities to completely deplete the initial molecular sample. For example, starting from $X(62, 0)$ molecules, we have found additional resonances when the laser frequencies are tuned to 413.923,36 THz, 414.726,41 THz, 415.493,19 THz, and 415.826,80 THz. Proper assignment of intermediately bound states and the exploration of $(2)0_u^+$ are left to future work.

As mentioned in Sect. 2.2, the $(1)0_u^+$ relativistic potential is mostly a superposition of $A^1\Sigma_u^+$ and $c^3\Pi_u$. The effect of the spin-orbit coupling is manifest in the binding energies of $(1)0_u^+$ at the avoided crossing. Figure 3.13 plots the binding energies of the deeply bound states with $J' = 1$ found by direct excitation in this work (black squares). As can be seen, there is a marked alteration in the trend of the binding energies near $-1500\,\text{cm}^{-1}$ with respect to the $^1S_0 + {}^3P_1$ threshold. For $v' \lesssim 15$, there is good agreement with the prescribed Dunham expansion from Ref. [30], where the potential parameters for $A^1\Sigma_g^+$ were determined from observations of transitions between deeply bound vibrational states of the ground and excited potentials with high angular momenta (J, $J' \sim 100$). For comparison, we also plot the binding energies computed from solving the Schrodinger equation using the Morse/Long-range potential and the ab initio potential, taken from Refs. [17] and [31] respectively. The location of (and the shape of the potential at) the crossing is challenging to model from first principles. As a result, the total number of vibrational states in $(1)0_u^+$ is 83 using the Morse/Long-range model, but 110 using the ab initio model.

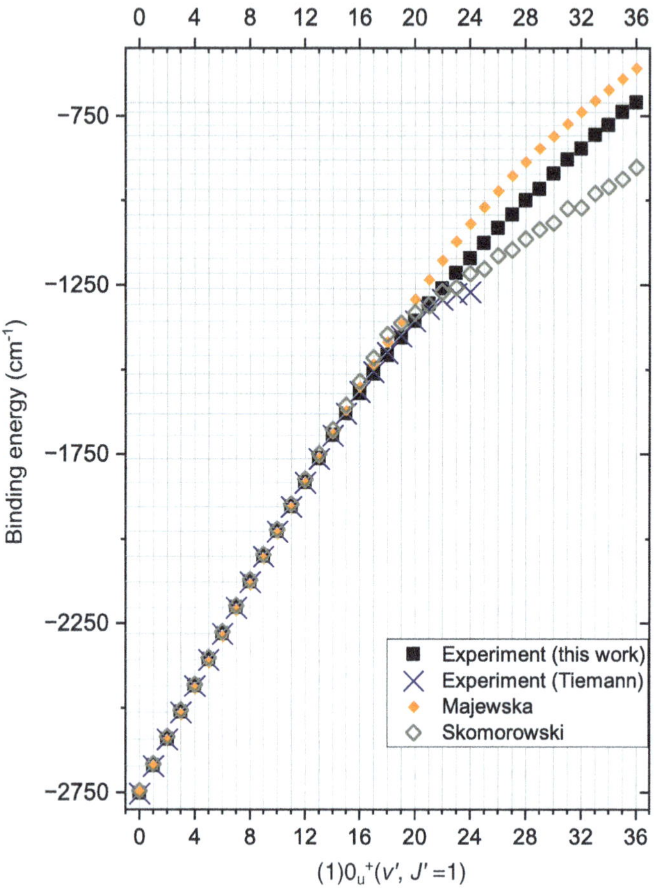

Fig. 3.13 Binding energies of deeply bound $(1)0_u^+$ states with $J' = 1$ relative to ${}^1S_0 + {}^3P_1$. Black squares, measured by direct one-photon excitation from $X(62, 0)$ as a result of this work and Ref. [23]. Purple crosses, extrapolated to $J' = 1$ states using the Dunham expansion and empirically determined (but isotopologue averaged) potential parameters for $A^1\Sigma_u^+$ from Ref. [30]. Orange diamonds, calculation using the Morse/Long-range potential from Ref. [17]. Gray open diamonds, ab initio calculation from Ref. [31]

3.5.3 $(1)1_u$

Deeply bound states of $(1)1_u$ with odd J', below the potential minimum of $(1)0_u^+$, are predominantly of $a^3\Sigma_u^+$ character and have not been observed before this thesis and Ref. [33]. A consolidation of all known $(1)1_u$ states so far is given in Table 3.6. Figure 3.14 plots the binding energies of deeply bound $(1)1_u$ states with $J' = 1$ measured in our experiment.

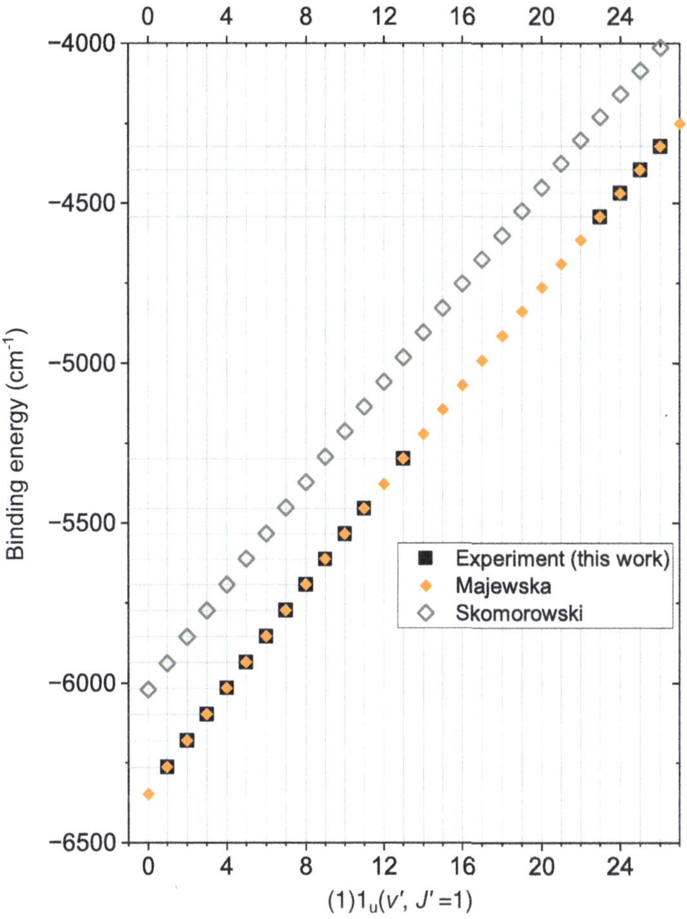

Fig. 3.14 Binding energies of deeply bound $(1)1_u$ states with $J' = 1$ relative to ${}^1S_0 + {}^3P_1$. Black squares, measured either by direct one-photon excitation from deeply bound $X^1\Sigma_g^+$ states or from dispersive lattice-driven resonances as a result of this work and Refs. [17, 23, 32]. Orange diamonds, calculation using the Morse/Long-range potential from Ref. [17]. Gray open diamonds, ab initio calculation from Ref. [31]

Owing to vanishing transition strengths from *weakly* bound $X^1\Sigma_g^+$ states, without an additional state preparation step (e.g., STIRAP to deeply bound $X^1\Sigma_g^+$ states), spectroscopy of the deeply bound, short-range part of $(1)1_u$ will have to be indirect. Indirect $X \rightarrow 1_u$ transitions are observed as lattice-driven polarizability resonances (see Sect. 5.3.2). Before Ref. [17], we used the ab initio model of Ref. [31] to assign the vibrational quantum numbers v' by matching the measured binding energies with the closest values calculated from the model despite substantial discrepancies on the order of $\sim 10\,\text{cm}^{-1}$, much greater than the typical error from the wavemeter instrument (HighFinesse, WS6-200). At the time, only seven deeply bound $(1)1_u$ levels with $J' = 1$ had been located. These were labeled as $v'_{\text{AI}} = 1$–3, 19–22, where the subscript "AI" explicitly reminds us that these assignments used the less accurate ab initio model and are therefore erroneous (Tables 3.3 and 3.4).

Another hint that the ab initio $(1)1_u$ model is inadequate in the short-range came from transition strength measurements, which we will discuss in Sect. 5.3.2. While the binding energy data were insufficient to determine the v' labels (unless the absolute ground vibrational state $v' = 0$ has been found), it is possible to figure out the *differences* in v'. Since $(1)1_u$ is unperturbed in the short-range, we model its behavior with the simple Morse potential [Eq. (3.47)]. Next, we take the difference of adjacent vibrational energy levels for $J' = 1$. Using Eq. (3.48), we get

$$E(v'+1, 1) - E(v', 1) = (\omega_e - 2\alpha_e) - 2\omega_e x_e (v'+1). \tag{3.50}$$

Therefore, plotting this difference against $v' + 1$ should yield a straight line with a slope equal to $-2\omega_e x_e$ that is insensitive to an overall offset in the v' labels [Fig. 3.15a]. The best fits, judged based on the reduced-χ^2 statistic, are obtained if there are 15 intermediate vibrational states in between the two groups. This implies that $\omega_e x_e = 0.21150(28)\,\text{cm}^{-1}$.

To determine α_e, we measure the rotational splitting of $J' = 1, 3$ for various v'. Using Eq. (3.48) for $|\Omega| = 1$, we see that the rotational splitting

$$E(v', 3) - E(v', 1) = 10B_e - 10\alpha_e \left(v' + \frac{1}{2}\right), \tag{3.51}$$

so that plotted against $v' + \frac{1}{2}$ the slope of the linear fit is $-10\alpha_e$ and insensitive to an overall offset in the v' labels. To attain the required accuracy and precision, we perform one-photon spectroscopy of $J' = 1, 3$ states using an additional laser on a sample of deeply bound $J = 2$ ground state molecules;[38] specifically $X(6, J = 2)$ for this study. We use short probe pulses ($\sim 100\,\mu\text{s}$) and low probe intensities to avoid bleaching the signal or power broadening the peaks. We linearly extrapolate the line centers to zero lattice power and determine the $X \rightarrow 1_u$ resonance frequencies and rotational splittings with the frequency comb. Table 3.6 lists the measured rotational splittings, and Fig. 3.15b shows the linear fit, from which we extract $\alpha_e = 7.068(11) \times 10^{-5}\,\text{cm}^{-1}$.

---

[38] In the work of Ref. [17], we had not implemented our STIRAP transfer technology. So $X(6, 2)$ molecules were inefficiently prepared with a long stimulated Raman pulse from $X(62, 0)$ in a non-magic lattice. We complete a roundtrip to transfer the population back to $X(62, 0)$ for detection.

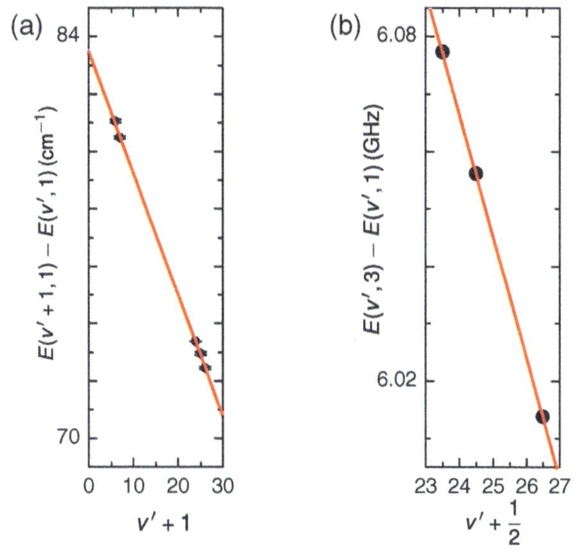

Fig. 3.15 (**a**) Adjacent vibrational splittings of $1_u(v', 1)$, plotted against $v' + 1$. (**b**) Rotational splittings between $J' = 3 - J' = 1$, plotted against $v' + \frac{1}{2}$. Solid red lines are linear fits to the data. The intercepts with the vertical axis are $\omega_e - 2\alpha_e$ and $10B_e$ respectively. At the time of Ref. [17], only the states shown in this figure were available. Subsequent experiments in this thesis have found more deeply bound $(1)1_u$ states, which are listed in Table 3.6

Next, we observe that for the linear fits in Fig. 3.15a and b, the vertical intercepts are $(\omega_e - 2\alpha_e)$ and $10B_e$, respectively, only *if* accurate knowledge of the v' labels are available. As we will see in Sect. 5.3.2, we can simultaneously reproduce the measured trends of the transition strengths from $X(6, 0)$ and the spectroscopic constants *only* if the vibrational quantum numbers are v' = 5–7, 23–26. With this at hand, we finally extract $\omega_e = 83.528(13)\,\text{cm}^{-1}$ and $B_e = 0.021933(3)\,\text{cm}^{-1}$. The equilibrium bond length is calculated using $R_e = \frac{1}{2\pi}\sqrt{h/(2\mu_2 c B_e)} = 7.9027(5)\,a_0$. Lastly, we note that the actual potential depth, D_e, relative to ${}^1S_0 + {}^3P_1$ will be overestimated by the standard formula $\omega_e^2/(4\omega_e x_e)$ for the simple Morse model because it does not extrapolate well to the long-range. Instead, we determine D_e by fitting the binding energies [Table 3.6] versus $v' + \frac{1}{2}$ to Eq. (3.48), with α_e, $\omega_e x_e$, and ω_e held fixed to the values above. The vertical intercept of the quadratic fit is $2B_e - D_e$. Knowing B_e, we find $D_e = 6387.89(11)\,\text{cm}^{-1}$.

Our measurements imply that the ab initio model underestimates D_e by approximately 300 cm^{-1} (5% relative difference); see also Fig. 3.16. This is confirmed in subsequent work (see Fig. 3.14 and Table 3.6) via the direct one-photon excitation of the predicted bound states $(1)1_u(v' = 1\text{–}4, 1)$ from samples of $X(0, 0)$ and $X(0, 2)$ created using STIRAP (see Chap. 4). Unfortunately, the rovibrational ground state $(1)1_u(0, 1)$ was ever-so-slightly out of reach; addressing the state from $X(0, 0)$ requires laser radiation at 1085 nm, which was not available in our lab at the time of the study.

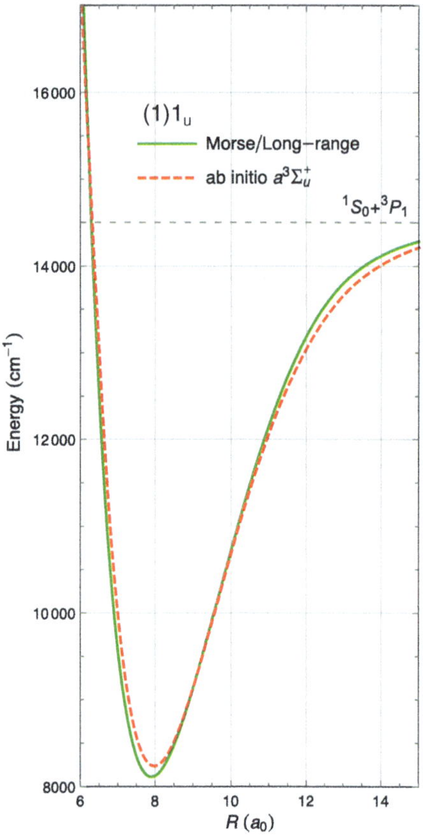

Fig. 3.16 Comparision of the ab initio [31] and Morse/Long-range [17] models for the $(1)1_u$ potential. The Morse/Long-range potential is more accurate, especially at short-range, near the potential minimum

3.5.4 Morse/Long-Range Potentials for $(1)1_u$ *and* $(1)0_u^+$

To combine the new insight about the short-range of $(1)1_u$ with the reliability of the ab initio calculation in the long-range, we recast both the ab initio $(1)1_u$ and $(1)0_u^+$ potentials in the Morse/Long-range (MLR) form [34, 35]. This is because a simultaneous fit of the long-range coefficients for both $(1)0_u^+$ and $(1)1_u$ potential curves is necessary to correctly describe the heavily Coriolis-mixed states $(1)0_u^+(-6, 1)$ and $(1)1_u(-3, 1)$.

A key strength of the Morse/Long-range form is its ability to accurately bridge between regions of a potential where data is unavailable. We fit[39] to the spectroscopic data shown in Tables 3.5 and 3.6, while fixing the depth D_e and the

[39] Note that at the time the fitting was performed, only the states listed in the Supplemental Material of Ref. [17] were available. A refit with newly available data in this thesis should allow the potential parameters to be more accurately determined (especially for $(1)0_u^+$ in the short- and intermediate-range), which we leave to future work.

equilibrium distance R_e of the $(1)1_u$ potential at their empirical values found in the previous subsection.

The Morse/Long-range (MLR) potential as a function of internuclear separation R has the form

$$V_{\mathrm{MLR}}(R) = D_e \left[1 - \frac{u_{\mathrm{LR}}(R)}{u_{\mathrm{LR}}(R_e)} e^{-\phi(R)\gamma_p(R)} \right]^2, \tag{3.52}$$

where

$$\begin{aligned} u_{\mathrm{LR}}(R) &= -\frac{C_3}{R^3} - \frac{C_6}{R^6} - \frac{C_8}{R^8} - \frac{C_{10}}{R^{10}}, \\ \phi(R) &= [1 - \gamma_p^{\mathrm{ref}}(R)] \sum_{i=0}^{N} (\gamma_q^{\mathrm{ref}}(R))^i \phi_i + \gamma_p^{\mathrm{ref}}(R)\phi_\infty, \\ \phi_\infty &= \ln\left(\frac{2D_e}{u_{\mathrm{LR}}(R_e)} \right), \\ \gamma_p(R) &= \frac{R^p - R_e^p}{R^p + R_e^p}, \\ \gamma_p^{\mathrm{ref}}(R) &= \frac{R^p - R_{\mathrm{ref}}^p}{R^p + R_{\mathrm{ref}}^p}. \end{aligned} \tag{3.53}$$

The initial conditions for fitting are the ab initio results from Skomorowski et al. [31]. Details regarding the fitting procedure can be found in Refs. [17, 36]. The resulting parameters are given in the upper half of Table 3.3. The spectroscopic constants for $(1)1_u$ are listed in the lower half of Table 3.3.

3.6 Spectroscopy Tables

In this section, we list the binding energies and rotational splittings of all known states in $^{88}Sr_2$ with $J = 0, 1, 2, 3, 4$, belonging to the potentials $X^1\Sigma_g^+$ (or $X0_g^+$), $(1)0_u^+$, and $(1)1_u$. All values are given in units of MHz, unless otherwise stated. A dash (–) indicates that the state has not yet been observed at the time of writing. A cross ($\times$) indicates that the state is not predicted to exist. Ref. [37] refers to this thesis. Note that "wrt." is shorthand for "with respect to".

To calculate the binding energies of the excited states in $(1)0_u^+$ and $(1)1_u$ relative to the $^1S_0 + {}^3P_1$ dissociation threshold in ^{88}Sr, we use the $^1S_0 \rightarrow {}^3P_1$ intercombination transition frequency from Refs. [39, 40], the binding energy of $X(62, 0)$ from Ref. [26], the vibrational splitting between $X(62, 0)$ and $X(61, 0)$ from Ref. [21], and the binding energies of various deeply bound $X^1\Sigma_g^+$ states given in Table 3.4.

Table 3.3 Morse/Long-range potential parameters of $(1)0_u^+$ and $(1)1_u$ in specified units. Also listed are the spectroscopic constants determined from deeply bound $(1)1_u$ states

	$(1)0_u^+$	$(1)1_u$
R_e (a_0)	7.5443	7.9027
D_e (cm^{-1})	2784	6388
C_3 ($E_h a_0^3$)	1.5235661×10^{-2}	7.6178307×10^{-3}
C_6 ($E_h a_0^6$)	3.8947894×10^3	4.0390241×10^3
C_8 ($E_h a_0^8$)	4.5157846×10^5	7.7660490×10^5
C_{10} ($E_h a_0^{10}$)	3.296×10^7	1.3253×10^8
p	9	9
q	4	4
R_{ref} (a_0)	8.2987	8.6930
ϕ_0	−0.63810976	−1.2454828
ϕ_1	3.5917033	−0.19418436
ϕ_2	7.7175691	−1.8890781
ϕ_3	0.57800325	−3.1121912
ϕ_4	−29.3700406	−6.0245946
ϕ_5	−23.080778	−5.6268047
ϕ_6	54.044018	−3.3425721
ϕ_7	91.862114	−0.0028626398
ϕ_8	35.061649	0
ϕ_9	−2.9283029	0
$\omega_e x_e$ (cm^{-1})	–	0.21150(28)
ω_e (cm^{-1})	–	83.528(13)
B_e (cm^{-1})	–	0.021933(3)
α_e (cm^{-1})	–	$7.068(11)\times 10^{-5}$

Table 3.4 $X^1\Sigma_g^+$ wrt. $^1S_0 + {}^1S_0$, values in units of MHz

$X0_g^+(v)$	$\lvert E_{J=0}\rvert$	$E_{J=2} - E_{J=0}$	Ref.
0	31,825,319.852(5)	3146(1)	[37]
1	30,640,296.437(28)	3125(10)	[37]
2	29,479,197.638(27)	3095(10)	[37]
3	28,342,002.355(26)	3060(10)	[37]
4	27,228,683.489(11)	3024(10)	[37]
5	26,139,208.322(26)	2986(10)	[37]
6	25,073,538.149(11)	2957(1)	[37]
7	24,031,629.148(26)	2925(10)	[37]
8	23,013,431.546(11)	2892(1)	[37]
9	22,018,900(120)	2860(10)	[37]
10	21,048,000(120)	2826(10)	[37]
11	20,100,500(120)	2782(10)	[37]
12	19,176,600(120)	2757(10)	[37]
13	18,276,000(120)	2722(10)	[37]
14	17,398,800(120)	2686(10)	[37]
15	16,544,800(120)	2653(10)	[37]
16	15,713,900(120)	2606(10)	[37]

(continued)

Table 3.4 (continued)

17	14,906,100(120)	2574(10)	[37]
18	14,121,300(120)	2552(10)	[37]
19	13,359,300(120)	2499(10)	[37]
20	12,620,100(120)	2469(10)	[37]
21	11,903,600(120)	2427(10)	[37]
22	11,209,700(120)	2386(10)	[37]
23	10,538,300(120)	2353(10)	[37]
24	9,889,287(120)	2309(10)	[37]
25	9,262,568(120)	2280(10)	[37]
26	8,658,009(120)	2230(10)	[37]
27	8,075,549(120)	2190(10)	[37]
28	7,515,067(120)	2149(10)	[37]
29	6,976,461(120)	2102(10)	[37]
30	6,459,628(120)	2059(10)	[37]
31	5,964,469(120)	2025(10)	[37]
32	5,490,845(120)	1967(10)	[37]
33	5,038,674(120)	1928(10)	[37]
34	4,607,839(120)	1890(10)	[37]
35	4,198,199(120)	1830(10)	[37]
36	3,809,659(120)	1790(10)	[37]
37	3,442,049(120)	1735(10)	[37]
38	3,095,249(120)	1690(10)	[37]
39	2,769,099(120)	1640(10)	[37]
40	2,463,424(120)	1585(10)	[37]
41	2,178,039(120)	1540(10)	[37]
42	1,912,719(120)	1480(10)	[37]
43	1,667,229(120)	1425(10)	[37]
44	1,441,289(120)	1365(10)	[37]
45	1,234,579(120)	1300(10)	[37]
46	1,046,739(120)	1250(10)	[37]
47	877,339(120)	1180(10)	[37]
48	725,889(120)	1120(10)	[37]
49	591,819(120)	1050(10)	[37]
50	474,464(120)	985(10)	[37]
51	373,094(120)	930(10)	[37]
52	286,789(120)	850(10)	[37]
53	214,639(120)	770(10)	[37]
54	155,564(120)	705(10)	[37]
55	108,369(120)	625(10)	[37]
56	71,824(120)	555(10)	[37]
57	44,599(120)	475(10)	[37]
58	25,339(120)	400(10)	[37]
59	12,639(120)	310(10)	[37]
60	5110.5739(51)	230(10)	[24, 26, 37]
61	1400.3182(50)	173.0(2)	[24, 26]
62	136.6447(50)	70.1(2)	[24, 26]

Table 3.5 $(1)0_u^+$ wrt. $^1S_0 + {}^3P_1$, values in units of MHz

$(1)0_u^+(v)$	$\lvert E_{J'=1}\rvert$	$E_{J'=3} - E_{J'=1}$	Ref.
0	82,462,408(60)	–	[23, 37]
1	80,056,883(60)	–	[23, 37]
2	77,669,758(60)	–	[23, 37]
3	75,301,073(60)	–	[23, 37]
4	72,950,983(60)	–	[23, 37]
5	70,619,778(60)	–	[23, 37]
6	68,307,898(60)	–	[23, 37]
7	66,016,088(60)	–	[23, 37]
8	63,745,468(60)	–	[23, 37]
9	61,497,833.94(12)	–	[23, 37]
10	59,275,910.26(35)	–	[23, 37]
11	57,084,156.51(12)	–	[23, 37]
12	54,929,909.90(11)	–	[23, 37]
13	52,825,520.41(4)	–	[23, 37]
14	50,791,292.56(10)	–	[23, 37]
15	48,855,512.13(18)	–	[23, 37]
16	47,036,433.95(23)	–	[23, 37]
17	45,320,332.03(14)	–	[23, 37]
18	43,686,942.35(9)	–	[23, 37]
19	42,124,784.98(7)	–	[23, 37]
20	40,655,281.22(8)	–	[23, 37]
21	39,174,338(60)	–	[23, 37]
22	37,763,228(60)	–	[23, 37]
23	36,377,628(60)	–	[23, 37]
24	35,066,118(60)	–	[23, 37]
25	33,737,818(60)	–	[23, 37]
26	32,397,458(60)	–	[23, 37]
27	31,241,318(60)	–	[23, 37]
28	29,959,413(60)	–	[23, 37]
29	28,947,228(60)	–	[23, 37]
30	27,621,073(60)	–	[23, 37]
31	26,361,623(60)	–	[23, 37]
32	25,400,438(60)	–	[23, 37]
33	24,186,203(60)	–	[23, 37]
34	23,295,548(60)	–	[23, 37]
35	22,096,778(60)	–	[23, 37]
36	21,284,598(60)	–	[23, 37]
37 to −7	–	–	–
−6	8429.650(42)	353.55(50)	[26, 38]
−5	3463.280(33)	284.18(50)	[26, 38]
−4	1084.093(33)	183.09(50)	[26, 38]
−3	222.161(35)	90.26(50)	[26, 38]
−2	23.9684(50)	23.342(13)	[24]
−1	0.4653(45)	×	[26]

Table 3.6 $(1)1_u$ wrt. $^1S_0 + {}^3P_1$, values in units of MHz

$(1)1_u(v)$	$\|E_{J'=1}\|$	$E_{J'=2} - E_{J'=1}$	$E_{J'=3} - E_{J'=1}$	$E_{J'=4} - E_{J'=1}$	Ref.
0	–	–	–	–	–
1	187,757,863(300)	–	–	–	[37]
2	185,281,896(300)	–	–	–	[37]
3	182,818,099(300)	–	–	–	[37]
4	180,366,282(300)	–	–	–	[37]
5	177,926,755(300)	–	–	–	[17, 37]
6	175,499,231(300)	–	–	–	[37]
7	173,083,903(300)	–	–	–	[17, 37]
8	170,681,023(300)	–	–	–	[37]
9	168,290,366(1)	2618(1)	6362(1)	–	[37]
10	165,912,119(300)	–	–	–	[37]
11	163,546,200(300)	–	–	–	[37]
12	–	–	–	–	–
13	158,851,898(300)	–	–	–	[37]
14 to 22	–	–	–	–	–
23	136,143,326(1)	–	6077.28(9)	–	[17, 33]
24	133,943,814(1)	2478(1)	6056.16(9)	–	[17, 33]
25	131,757,352(100)	–	–	–	[17, 33]
26	129,584,290(1)	2458(1)	6013.75(6)	–	[17, 33]
27 to −4	–	–	–	–	–
−3	8200.163(39)	87(1)	539.6(5)	733(1)	[26, 38]
−2	2683.722(32)	115(1)	328.5(5)	530(1)	[26, 38]
−1	353.236(35)	66(1)	182(1)	297(1)	[21, 26, 38]

Errata

- For $(1)1_u(6, 1)$, the value reported in Ref. [17] is erroneous due to a numerical rounding error. This has been revised in this thesis.
- For $(1)0_u^+(v, 1)$, $v = 13, 16, 18, 19$ and 20, the values reported in Ref. [23] are possibly erroneous due to inaccurate wavemeter calibration, leading to the misidentification of the comb teeth numbers. These have been revised in this thesis.

References

1. Legero, T., Kessler, T., Sterr, U.: Tuning the thermal expansion properties of optical reference cavities with fused silica mirrors. J. Opt. Soc. Am. B **27**, 914–919 (2010)
2. Liedl, C.: A molecular lattice clock. MA thesis (Ludwig Maximilian University of Munich, Munich, 2018, project at Columbia University)
3. Alnis, J., Matveev, A., Kolachevsky, N., Udem, T., Hänsch, T.: Subhertz linewidth diode lasers by stabilization to vibrationally and thermally compensated ultralow-expansion glass Fabry-Pérot cavities. Phys. Rev. A **77**, 053809 (2008)

4. Fox, R.W.: Temperature analysis of low-expansion Fabry-Perot cavities. Opt. Express **17**, 15023–15031 (2009)
5. Fox, R.W., Oates, C.W., Hollberg, L.W.: Experimental Methods in the Physical Sciences, pp. 1–46. Elsevier, Amsterdam (2003)
6. Black, E.D.: An introduction to Pound–Drever–Hall laser frequency stabilization. Am. J. Phys. **69**, 79–87 (2001)
7. McGuyer, B.H., et al.: Precise study of asymptotic physics with subradiant ultracold molecules. Nat. Phys. **11**, 32–36 (2015)
8. Shen, H., Li, L., Bi, J., Wang, J., Chen, L.: Systematic and quantitative analysis of residual amplitude modulation in Pound-Drever-Hall frequency stabilization. Phys. Rev. A **92**, 063809 (2015)
9. Wong, N., Hall, J.L.: Servo control of amplitude modulation in frequency-modulation spectroscopy: demonstration of shot-noise-limited detection. J. Opt. Soc. Am. B **2**, 1527–1533 (1985)
10. Zhang, W., et al.: Reduction of residual amplitude modulation to 1×10^{-6} for frequency modulation and laser stabilization. Opt. Lett. **39**, 1980–1983 (2014)
11. Fortier, T., Baumann, E.: 20 years of developments in optical frequency comb technology and applications. Commun. Phys. **2**, 1–16 (2019)
12. Cundiff, S.T., Ye, J.: Colloquium: femtosecond optical frequency combs. Rev. Mod. Phys. **75**, 325 (2003)
13. Marlow, B.L.S., Scherer, D.R.: A review of commercial and emerging atomic frequency standards. IEEE Trans. Ultrason. Ferroelectr. Freq. Control **68**, 2007–2022 (2021)
14. Benkler, E., Lisdat, C., Sterr, U.: On the relation between uncertainties of weighted frequency averages and the various types of Allan deviations. Metrologia **52**, 565 (2015)
15. Dawkins, S.T., McFerran, J.J., Luiten, A.N.: Considerations on the measurement of the stability of oscillators with frequency counters. IEEE Trans. Ultrason. Ferroelectr. Freq. Control **54**, 918–925 (2007)
16. Rubiola, E.: On the measurement of frequency and of its sample variance with high-resolution counters. Rev. Sci. Instrum. **76**, 054703 (2005)
17. Leung, K.H., et al.: Transition strength measurements to guide magic wavelength selection in optically trapped molecules. Phys. Rev. Lett. **125**, 153001 (2020)
18. Telle, H.R., Lipphardt, B., Stenger, J.: Kerr-lens, mode-locked lasers as transfer oscillators for optical frequency measurements. Appl. Phys. B **74**, 1–6 (2002)
19. Stenger, J., Schnatz, H., Tamm, C., Telle, H.R.: Ultraprecise measurement of optical frequency ratios. Phys. Rev. Lett. **88**, 073601 (2002)
20. Brochard, P., Schilt, S., Südmeyer, T.: Ultra-low noise microwave generation with a free-running optical frequency comb transfer oscillator. Opt. Lett. **43**, 4651–4654 (2018)
21. McGuyer, B., et al.: Control of optical transitions with magnetic fields in weakly bound molecules. Phys. Rev. Lett. **115**, 053001 (2015)
22. Jackson, S., Vutha, A.C.: Magic polarization for cancellation of light shifts in two-photon optical clocks. Phys. Rev. A **99**, 063422 (2019)
23. Leung, K.H., et al.: Ultracold $^{88}Sr_2$ molecules in the absolute ground state. New J. Phys. **23**, 115002 (2021)
24. McGuyer, B., et al.: High-precision spectroscopy of ultracold molecules in an optical lattice. New J. Phys. **17**, 055004 (2015)
25. Reinaudi, G., Osborn, C., McDonald, M., Kotochigova, S., Zelevinsky, T.: Optical production of stable ultracold $^{88}Sr_2$ molecules. Phys. Rev. Lett. **109**, 115303 (2012)
26. McDonald, M.: High Precision Optical Spectroscopy and Quantum State Selected Photodissociation of Ultracold $^{88}Sr_2$ Molecules in an Optical Lattice. Springer, Berlin (2017)
27. Stein, A., Knöckel, H., Tiemann, E.: Fourier-transform spectroscopy of Sr_2 and revised ground-state potential. Phys. Rev. A **78**, 042508 (2008)
28. Stein, A., Knöckel, H., Tiemann, E.: The $^1S + {}^1S$ asymptote of Sr_2 studied by Fourier-transform spectroscopy. Eur. Phys. J. D **57**, 171–177 (2010)

29. McGuyer, B., et al.: Nonadiabatic effects in ultracold molecules via anomalous linear and quadratic Zeeman shifts. Phys. Rev. Lett. **111**, 243003 (2013)
30. Stein, A., Knöckel, H., Tiemann, E.: The states $1^1\Sigma_u^+$, $1^1\Pi_u$ and $2^1\Sigma_u^+$ of Sr_2 studied by Fourier-transform spectroscopy. Eur. Phys. J. D **64**, 227–238 (2011)
31. Skomorowski, W., Pawłowski, F., Koch, C.P., Moszynski, R.: Rovibrational dynamics of the strontium molecule in the $A^1\Sigma_u^+$, $c^3\Pi_u$, and $a^3\Sigma_u^+$ manifold from state-of-the-art ab initio calculations. J. Chem. Phys. **136**, 194306 (2012)
32. Kondov, S.S., et al.: Molecular lattice clock with long vibrational coherence. Nat. Phys. **15**, 1118–1122 (2019)
33. Lee, C.-H.: Quantum metrology with a molecular lattice clock and state-selected photodissociation of ultracold molecules. PhD thesis (Columbia University, New York, 2020)
34. Le Roy, R.J., et al.: Accurate analytic potentials for Li_2 ($X^1\Sigma_g^+$) and Li_2 ($A^1\Sigma_u^+$) from 2 to 90 Å, and the radiative lifetime of Li($2p$). J. Chem. Phys. **131**, 204309 (2009)
35. Le Roy, R.J., Henderson, R.D.: A new potential function form incorporating extended long-range behaviour: application to ground-state Ca_2. Mol. Phys. **105**, 663–677 (2007)
36. Majewska, I.: Theoretical description of ultracold strontium molecules in an optical lattice: control of photodissociation and interpretation of molecular clock experiments. PhD thesis (University of Warsaw, Warsaw, 2021)
37. Leung, K.H.: The strontium molecular lattice clock: vibrational spectroscopy with hertz-level accuracy. PhD thesis (Columbia University, New York, 2023)
38. Zelevinsky, T., et al.: Narrow line photoassociation in an optical lattice. Phys. Rev. Lett. **96**, 203201 (2006)
39. Ido, T., et al.: Precision spectroscopy and density-dependent frequency shifts in ultracold Sr. Phys. Rev. Lett. **94**, 153001 (2005)
40. Ferrari, G., et al.: Precision frequency measurement of visible intercombination lines of strontium. Phys. Rev. Lett. **91**, 243002 (2003)

Chapter 4
Ultracold $^{88}Sr_2$ Molecules in the Absolute Ground State

4.1 An Intuitive Description of Stimulated Raman Adiabatic Passage (STIRAP)

Stimulated Raman adiabatic passage (STIRAP) is an ingenious and powerful coherent state transfer technique, finding numerous applications in widely disparate fields [1, 2], including ultracold molecules, atomtronics, quantum computing, solid-state physics, atmospheric chemistry, etc. Under favorable experimental conditions, STIRAP transfer efficiencies between the initial and target quantum states can be nearly 100%. A comprehensive account and theoretical overview of STIRAP is given in Refs. [3, 4]. STIRAP is superficially similar to adiabatic rapid passage (ARP); both are laser-based methods for highly efficient quantum state transfer. But while ARP is conceptually based on Landau-Zener theory (where the frequency of the laser is adiabatically swept across resonance), STIRAP is more closely related to the phenomenon of dark states.

Rather than belaboring the theoretical formalism for STIRAP, which is readily found in literature and textbooks, it would be beneficial to gain an intuitive understanding of its salient features. Consider a three-level system in the Λ-configuration [Fig. 4.1a]. Laser fields couple two metastable ground states ($|g_1, g_2\rangle$) to a third excited state ($|e\rangle$), but not to each other. The fields induce two distinct oscillating dipoles that destructively (constructive) interfere, forming a dark (bright) state that has a suppressed (an enhanced) outgoing radiation rate. For simplicity, suppose the laser polarizations are parallel to the quantization axis (lab frame $\hat{\mathbf{Z}}$-axis), so that the angular Rabi frequencies [Eqs. (3.9) and (3.15)] of the pump and anti-Stokes[1] are $\Omega_1 = \langle e|d_0^{(1)}|g_1\rangle \mathcal{E}_{0,1}/\hbar$ and $\Omega_2 = \langle e|d_0^{(1)}|g_2\rangle \mathcal{E}_{0,2}/\hbar$. The dark state

[1] In the context of STIRAP, it is common to refer to the upleg as the "pump" and the downleg as the "(anti-)Stokes", if the latter has a (shorter) longer wavelength than the upleg. In the context of two-photon spectroscopy and coherent population trapping (CPT) clocks, it is also common to see the

K. H. Leung, *The Strontium Molecular Lattice Clock*, Springer Theses,
https://doi.org/10.1007/978-3-031-47647-1_4

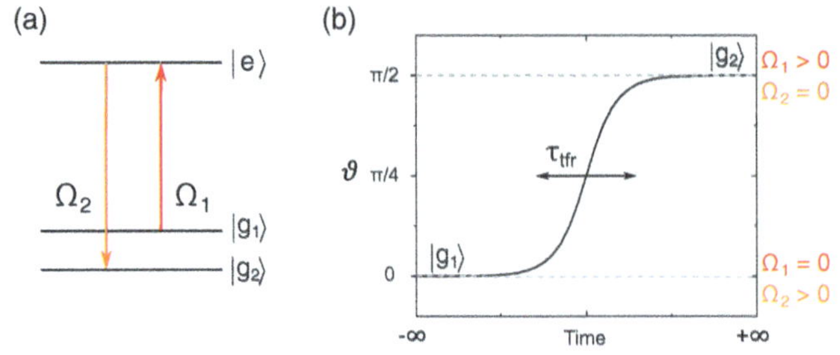

Fig. 4.1 Illustration of the principle of coherent population transfer with STIRAP. (**a**) A three-level system, in a Λ-configuration coupled by two laser fields, exhibits a single dark state that is the coherent superposition of the ground states with amplitudes proportional to the Rabi frequencies. (**b**) Time evolution of the mixing angle during a STIRAP sequence. The amplitudes of the dark state track the adiabatic evolution of the Hamiltonian when the Rabi frequencies are varied by modulating the laser powers of the pump and anti-Stokes, within a transfer time τ_{tfr} that is longer than the time scale of the ordinarily driven dynamics ($\sim 1/\Omega_{\mathrm{rms}}$)

may be expressed as the superposition $|D\rangle = \Omega_2|g_1\rangle - \Omega_1|g_2\rangle$, which can be verified in a single algebraic step; i.e., the matrix element of the perturbing Hamiltonian[2] $H'_{\mathrm{rwa}} = d_0^{(1)}\mathcal{E}_{0,1} + d_0^{(1)}\mathcal{E}_{0,2}$ between the dark and excited state is $\langle e|H'_{\mathrm{rwa}}|D\rangle = 0$, as expected.

Generally speaking, for a multi-level system coupled by multiple electromagnetic fields, the number of dark states is at least $n_g - n_e$, where n_g (n_e) is the number of ground (excited) states interacting with light [5, 6]. In the current discussion, $n_g = 2$ and $n_e = 1$, so we have exactly one dark state. Suppose the molecules are initialized in state $|g_1\rangle$, with the downleg maximally irradiating the molecules and the upleg completely shuttered off. The downleg would not deplete any population in $|g_1\rangle$, so $|g_1\rangle$ is the dark state. As a sanity check, substituting $\Omega_2 \neq 0$ and $\Omega_1 = 0$ into $|D\rangle$, we indeed obtain (after normalization) $|D\rangle = |g_1\rangle$. By symmetry, if the downleg is completely shuttered off and the upleg maximally irradiates the molecules, then the dark state is $|D\rangle = |g_2\rangle$. Just like the notorious game of Three-card Monte, we perform the sleight of hand by fully ramping down Ω_2 while simultaneously ramping up Ω_1, leaving the molecules "in the dark" (i.e., dark state) as we "shuffle the money card" (i.e., ground molecular population) under the hood. In more concrete physical terms, we achieve complete population transfer by adiabatically evolving the light-molecule Hamiltonian, which in turn coherently transforms (rotates) the dark state from $|g_1\rangle$ to $|g_2\rangle$,

upleg referred to as the "probe" and the downleg as the "coupling". All three sets of terminology are widely used in literature.

[2] In the rotating wave approximation, and ignoring the overall minus sign.

$$|D(t)\rangle = \Omega_2(t)|g_1\rangle - \Omega_1(t)|g_2\rangle = \cos\vartheta(t)\,|g_1\rangle - \sin\vartheta(t)\,|g_2\rangle = \begin{cases} |g_1\rangle, & t = -\infty, \\ |g_2\rangle, & t = +\infty, \end{cases} \tag{4.1}$$

where the mixing angle is defined as $\tan\vartheta(t) \equiv \Omega_1(t)/\Omega_2(t)$ and we have ignored normalization constants. Figure 4.1b illustrates this process. As the population is kept in $|D\rangle$ throughout the entire duration of the sequence, scattering (or depletion) losses due to the interrogating laser beams are eliminated, which would otherwise diminish the transfer efficiency. This is the key principle behind STIRAP.

To maintain the **adiabaticity condition**, a rule-of-thumb is that $\Omega_{\mathrm{rms}}\tau_{\mathrm{tfr}} > 10$, where $\Omega_{\mathrm{rms}} \equiv \sqrt{\Omega_1^2 + \Omega_2^2}$ and τ_{tfr} is the STIRAP transfer time [3]. In practice, the availability of laser power limits the respective Rabi frequencies, setting a minimum duration for τ_{tfr}. The upleg tends to be the bottleneck for ultracold molecules, requiring large laser intensities due to poorer Frank-Condon overlap. This has to be weighed against the relative coherence of the laser fields; i.e., the relative coherence time should be longer than τ_{tfr} so that the dark state remains robust during the STIRAP transfer. It can be shown [7, 8] that the idealized one-way STIRAP efficiency is $\sim \exp\left(-\frac{\pi^2\Gamma}{\Omega_{\mathrm{rms}}^2\tau_{\mathrm{tfr}}}\right)$. Fortunately, laser sources in the visible and near-IR spectrum have gotten so reliable and straightforward to phase stabilize (either with a common reference cavity or a frequency comb) that STIRAP is considered a routine step for many experiments working with ultracold molecules. The intrinsically higher amplified spontaneous emission (ASE) and phase noise spectrum in diode laser systems can be detrimental, especially when the ground or excited manifold of sub-states are densely spaced or unresolved [9–11]. Ways to circumvent or mitigate these deleterious effects include performing STIRAP at a finite common detuning from the excited state manifold, and filtering the laser light using modestly high finesse optical cavities [8, 12].

4.2 Finding a Two-Photon Pathway

To identify a suitable two-photon STIRAP pathway, we must survey the excited state structure and their transition strengths to the ground states. From the general considerations above, we see that to achieve high-efficiency transfer within the singlet $X^1\Sigma_g^+$ ground potential, we should seek excited states with marginal triplet admixture and favorable Frank-Condon overlap with the initial and final states, which maximizes the achievable Rabi frequencies. Additionally, while STIRAP is theoretically agnostic to the excited state natural linewidth (Γ) if the dark state is perfectly created, it would be an excellent strategy to pick excited states that have reasonable linewidths (i.e., *not* ~GHz) to anticipate imperfections in the state coupling that inevitably arise in actual experiments.

4.2.1 Molecular Transition Strengths (General Discussion)

Consider the initial state $|\alpha;\, J, M\rangle$ that **absorbs** a photon from the field

$$\boldsymbol{\mathcal{E}} = \frac{\mathcal{E}_0}{2}\left(\boldsymbol{\epsilon}\, e^{-i\omega t} + \boldsymbol{\epsilon}^*\, e^{+i\omega t}\right)$$

to make a transition to the energetically excited final state $|\beta;\, J', M'\rangle$. Here, α and β are just generic state labels. Using the expression $I = \epsilon_0 c \mathcal{E}_0^2/2$ for the irradiance [Eq. (2.15)], the angular Rabi frequency (Ω_R) characterizing the strength of coupling between two states due to the applied field is [Eq. (3.9)]

$$\Omega_R \equiv \frac{\mathcal{E}_0}{\hbar}\langle\beta;\, J', M'|\, \mathbf{d}\cdot\boldsymbol{\epsilon}\, |\alpha;\, J, M\rangle. \tag{4.2}$$

Now suppose that the laser polarization is fully polarized in a particular spherical component p to drive a specific transition from M to M'; i.e., ϵ_p is zero except for one p. Then taking the absolute square, we write

$$|\Omega_R|^2 = \frac{\mathcal{E}_0^2}{\hbar^2} S = \frac{2I}{\hbar^2 \epsilon_0 c} S, \tag{4.3}$$

where the transition (line) strength S is defined as

$$S \equiv \left|\langle\beta;\, J', M'|d^{(1)}_{M'-M}|\alpha;\, J, M\rangle\right|^2 = \begin{pmatrix} J' & 1 & J \\ -M' & M'-M & M \end{pmatrix}^2 \left|\langle\beta(J')||d^{(1)}||\alpha(J)\rangle\right|^2. \tag{4.4}$$

In the second step we used[3] the Wigner-Eckart theorem [Eq. (3.13)]. Colloquially, we call S the "DMS"; i.e., dipole matrix-element *squared.* Following this line of thought, for a molecular transition from $|n, v;\, J, M, \Omega\rangle$ to $|n', v';\, J', M', \Omega'\rangle$, where n denotes the electronic channel, and v the vibrational quantum number, the transition (line) strength is

$$S = \left|H^{JM\Omega}_{J'M'\Omega'}\right|^2 \times \left|\langle n', v'|\tilde{d}^{(1)}_{\Omega'-\Omega}(R)|n, v\rangle\right|^2, \tag{4.5}$$

where $\tilde{d}^{(1)}_{\Omega'-\Omega}(R)$ is the **electronic transition dipole moment in the molecule-fixed frame** dependent on the internuclear distance R, and we used the Born-Oppenheimer approximation to write $|n, v;\, J, M, \Omega\rangle \approx |n, v\rangle|J, M, \Omega\rangle$. The spatial wavefunction of the vibronic state $|n, v\rangle$ implicitly depends on J, Ω due to

[3] Some authors reserve double-barred matrix elements only in the molecule(body)-fixed frame. This thesis follows this convention very loosely and uses double-barred matrix elements whenever we invoke the Wigner-Eckart theorem.

centrifugal distortion. Multichannel wavefunctions can be accounted for by taking the sum over n' before evaluating the absolute square. The **rotational factor** is defined as

$$\begin{aligned} H_{J'M'\Omega'}^{JM\Omega} \equiv& (-1)^{M-\Omega}\sqrt{(2J+1)(2J'+1)} \\ &\times \sqrt{1+\delta^{\mathrm{kr}}_{\Omega,0}+\delta^{\mathrm{kr}}_{\Omega',0}-2\delta^{\mathrm{kr}}_{\Omega,0}\delta^{\mathrm{kr}}_{\Omega',0}} \\ &\times \begin{pmatrix} J' & 1 & J \\ -M' & M'-M & M \end{pmatrix} \begin{pmatrix} J' & 1 & J \\ -\Omega' & \Omega'-\Omega & \Omega \end{pmatrix}, \end{aligned} \tag{4.6}$$

where $\delta^{\mathrm{kr}}_{i,j}$ is the Kronecker delta. Equation (4.6) comes from rotating the lab-frame moment $d^{(1)}$ into the molecule-frame moment $\tilde{d}^{(1)}$, and evaluating the integral of a triple product of Wigner D-functions;[4] see Ref. [13] for the full expression of the integral, and other useful formulas. In literature, the sum over all sub-levels $\sum_{M,M'} |H_{J'M'\Omega'}^{JM\Omega}|^2$ is called the Hönl-London factor [14]. For $J = 0 \rightarrow J' = 1$, we do not take this sum as we have sufficient control over laser polarization and frequency to address individual sub-levels. Note that $|H_{1M'0}^{000}|^2 = 1/3$ for parallel transitions ($\Delta\Omega = 0$, e.g., $X \rightarrow 0_u^+$), and $|H_{1M'1}^{000}|^2 = 2/3$ for perpendicular transitions ($\Delta\Omega = 1$, e.g., $X \rightarrow 1_u$).

We can further expand the electronic moment in the molecule-fixed frame as a Taylor series (usually about the equilibrium distance, R_e),

$$\tilde{d}^{(1)}_{\Omega'-\Omega}(R) \approx \tilde{d}^{(1)}_{\Omega'-\Omega}(R_e) + (R-R_e)\left.\frac{\mathrm{d}(\tilde{d}^{(1)}_{\Omega'-\Omega})}{\mathrm{d}R}\right|_{R_e} + \ldots. \tag{4.7}$$

Next, applying the Born-Oppenheimer approximation to the vibronic states $|n, v\rangle \approx |v\rangle|n\rangle$, the matrix element can be expanded as

$$\begin{aligned} \langle n', v'|\tilde{d}^{(1)}_{\Omega'-\Omega}(R)|n, v\rangle \approx& \langle v'|v\rangle \times \langle n'|\tilde{d}^{(1)}_{\Omega'-\Omega}(R_e)|n\rangle \\ &+ \langle n', v'|\left[(R-R_e)\left.\frac{\mathrm{d}(\tilde{d}^{(1)}_{\Omega'-\Omega})}{\mathrm{d}R}\right|_{R_e}\right]|n, v\rangle \\ &+ \ldots. \end{aligned} \tag{4.8}$$

Taking the absolute square, the zeroth-order term is proportional to the **Franck-Condon factor**,

$$\mathrm{FCF} \equiv \left|\langle v'|v\rangle\right|^2, \tag{4.9}$$

[4] Recall from Sect. 2.2 that the Wigner D-functions represent the rotational wavefunctions of a symmetric top.

and an "electronic factor" $|\langle n'|\tilde{d}^{(1)}_{\Omega'-\Omega}(R_e)|n\rangle|^2$. We reiterate that the spatial wavefunctions of $|v\rangle$, $|v'\rangle$, $|n\rangle$, $|n'\rangle$ all implicitly depend on J, Ω, J', Ω' through centrifugal distortion. Returning to Eq. (4.5), we may intuitively understand molecular transition strengths in terms of their various degrees of freedom,

$$S \sim |\text{Rotational factor}|^2 \times \text{FCF} \times \text{Electronic factor}. \tag{4.10}$$

If $S = 0$, the transition is said to be *forbidden*.

4.2.2 *X to* (1)0_u^+ *Transition Strength Measurements*

In an earlier work by our theory collaborators [15, 16], deeply bound (1)0_u^+ states in the vicinity of the $A^1\Sigma_u^+$-$c^3\Pi_u$ avoided crossing were predicted to have favorable properties as intermediate states for a two-photon transfer between states of the ground potential $X^1\Sigma_g^+$. As we shall see, (1)0_u^+ transitions strengths with deeply bound ground state molecules can be as large as $10^{-2}(ea_0)^2$, while simultaneously maintaining reasonable strengths with weakly bound, near-threshold ground state molecules. Figure 4.2a, b plot the transition strengths of the two weakest bound ground vibrational states to (1)$0_u^+(v', J' = 1)$ states in the range $v' = 9$–20. The values are also listed in Table 4.1. These shallow-to-deep $X \rightarrow 0_u^+$ transition strengths are extracted from the one-photon spectra and excitation rates, as delineated in Sect. 3.3. To determine the irradiance [Eq. (2.15)], we estimate the focused upleg laser beam waist at the position of the molecules by deflecting the laser beam just before the vacuum chamber window onto an inexpensive CMOS or CCD camera. In general, the direct determination of beam waists can be tricky and susceptible to systematic errors, hence we assign conservative uncertainties.

As can be seen in Fig. 4.2a, our data suggests that (1)$0_u^+(11, 1)$ offers one of the strongest pump couplings for $X(62, 0)$. Moreover, the required laser wavelengths of 793 nm and 732 nm (for the pump and anti-Stokes, respectively) to address the entire depth of $X^1\Sigma_g^+$ are within the operating range of commercially available AR-coated laser diodes, and our frequency comb. While transitions starting from $X(62, 0)$ are generally weaker than those from the adjacently bound $X(61, 0)$, this is outweighed by our ability to create the former in larger samples and more efficiently detect them. Thus, for our proof-of-principle experiment, we choose $X(62, 0)$ as our initial state and the singlet-dominant (1)$0_u^+(11, 1)$ as the intermediate state for the STIRAP transfer.

Next, we lock the anti-Stokes laser on resonance with the absolute rovibrational ground state of the ^{88}Sr$_2$ dimer; i.e., $X(0, 0) \rightarrow$ (1)$0_u^+(11, 1)$. Keeping the anti-Stokes power low ($|\Omega_2| \ll \Gamma$) and using pulse durations within the relative coherence time, a scan of the pump across the (1)0_u^+ resonance reveals a narrow EIT peak (see Fig. 3.10a and Sect. 3.4.2), heralding the formation of a dark state necessary for STIRAP transfer. Using Autler-Townes spectroscopy [Fig. 3.10b–f], we

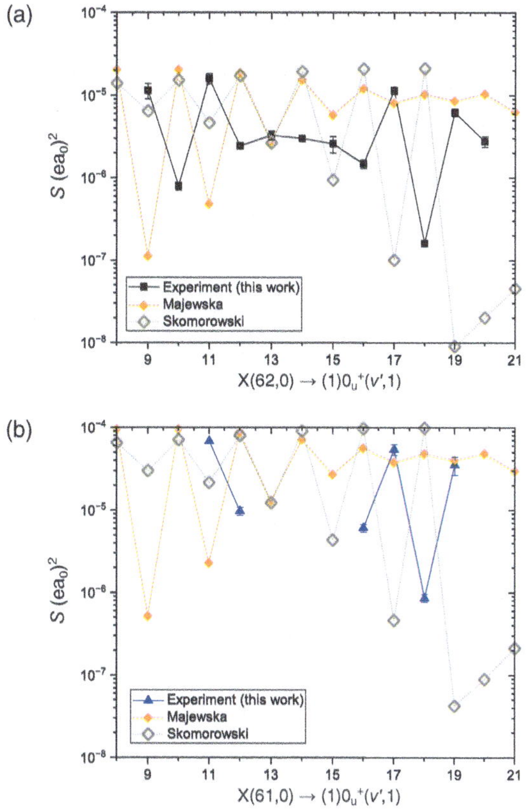

Fig. 4.2 Deeply bound $(1)0_u^+$ transition strengths from the weakly bound near-threshold states, (**a**) $X(62, 0)$ (black squares) and (**b**) $X(61, 0)$ (blue triangles). In both plots, yellow diamonds are computed using the Morse/Long-range model from Ref. [17], and open gray diamonds from Ref. [18]. The highly oscillatory behavior here is attributed to the alternating admixture of singlet and triplet components in the $(1)0_u^+$ states. The seeming lack of agreement between the experiment and the theoretical models is likely due to the difficulty in predicting the avoided crossing of $A^1\Sigma_u^+$ and $c^3\Pi_u$

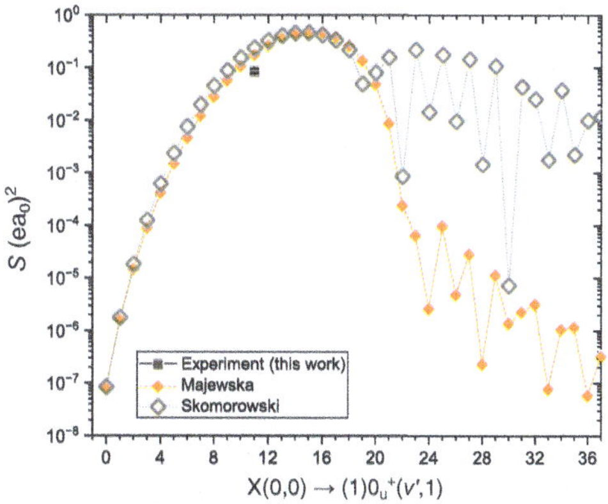

Fig. 4.3 Deeply bound $(1)0_u^+$ transition strengths from the absolute ground state $X(0, 0)$. Yellow diamonds are computed using the Morse/Long-range potential from Ref. [17], and open gray diamonds from Ref. [18]

measure a transition strength of $8.6(9) \times 10^{-2}(ea_0)^2$ for $X(0, 0) \rightarrow (1)0_u^+(11, 1)$. As shown in Fig. 4.3, this is one of the strongest molecular lines in $^{88}Sr_2$, with a value approaching that of a typical atomic transition. Intuitively, this behavior is attributed to the large $A^1\Sigma_u^+$ component of deeply bound $(1)0_u^+$ states, $v' \lesssim 20$.

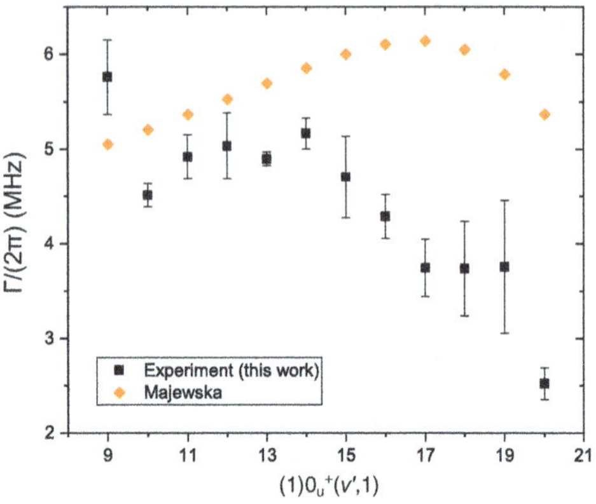

Fig. 4.4 Deeply bound $(1)0_u^+$ natural linewidths. Black squares, measured; yellow diamonds, computed using the Morse/Long-range potential from Ref. [17]

We have similarly measured the transition strengths from weakly bound $X^1\Sigma_g^+$ states to weakly bound $(1)0_u^+$ states using Autler-Townes spectroscopy [Table 4.2]. While there is a good agreement for these shallow-to-shallow transition strengths between our measurements and both models by our theory collaborators (*ab initio* [18], MLR [17]), notably only the MLR model gives the correct 0_u^+ or 1_u assignments for the pair of strongly Coriolis-mixed states bound by $\approx$8.3 GHz. One drawback of the MLR model is its omission of the $(2)0_u^+$ potential, which the *ab initio* model takes into account. As mentioned in Sect. 2.2 and shown in Fig. 2.2b, $(2)0_u^+$ extends below the intercombination threshold and affects transition strengths from deeply bound ground states to weakly bound $(1)0_u^+$ states. Weakly bound $(1)0_u^+$ states generally do not facilitate transfer ground state transfer except in fortuitous circumstances; e.g., $X(62, 0) \rightarrow X(4, 0)$ via $(1)0_u^+(-5, 1)$ as in Ref. [17], and $X(62, 0) \rightarrow X(6, 0)$ via $(1)0_u^+(-4, 1)$ as in Ref. [19]. Experimentally, the procedure for measuring deep-to-shallow $X \rightarrow 0_u^+$ transition strengths is the same (i.e., Autler-Townes spectroscopy), and the results are given in the lower part of Table 4.2 for the sake of completeness.

At bias magnetic fields of < 1 G the Zeeman sub-levels of $(1)0_u^+(11, 1)$ are essentially unresolved due to the $\sim$5 MHz excited state linewidth [Fig. 4.4]. The natural linewidths (Γ) of deeply bound $(1)0_u^+$ states are nearly 2 orders of magnitude larger than those of weakly bound $(1)0_u^+$ states, again because of the large $A^1\Sigma_u^+$ admixture.

4.3 Creation of $X(0, 0)$ Molecules Using STIRAP

To realize STIRAP in the laboratory, we need to dynamically change the intensity of the pump and anti-Stokes over typically $\sim$10 μs durations. To this end, each STIRAP laser beam is double-passed through an acousto-optic modulator (AOM). The amplitude of the microwave frequency driving the AOM is deterministically shaped

Table 4.1 Transition strengths of deeply bound $(1)0_u^+(J' = 1)$ with weakly and deeply bound $X^1\Sigma_g^+(J = 0)$, given in units of $(ea_0)^2$. Values under the header "Skomorowski" are calculated using the unscaled *ab initio* potential [18], while those under "Majewska" are calculated using the Morse/Long-range potential [17]

$X(v, 0)$	$(1)0_u^+(v', 1)$	Skomorowski	Majewska	Exp. (this work)
62	9	6.6×10^{-6}	1.2×10^{-7}	$1.1(2) \times 10^{-5}$
62	10	1.6×10^{-5}	2.1×10^{-5}	$8.0(8) \times 10^{-7}$
62	11	4.7×10^{-6}	5.0×10^{-7}	$1.6(2) \times 10^{-5}$
62	12	1.8×10^{-5}	1.9×10^{-5}	$2.4(2) \times 10^{-6}$
62	13	2.7×10^{-6}	2.9×10^{-6}	$3.3(4) \times 10^{-6}$
62	14	2.0×10^{-5}	1.6×10^{-5}	$3.0(2) \times 10^{-6}$
62	15	9.6×10^{-7}	6.1×10^{-6}	$2.6(6) \times 10^{-6}$
62	16	2.1×10^{-5}	1.3×10^{-5}	$1.5(2) \times 10^{-6}$
62	17	1.0×10^{-7}	8.4×10^{-6}	$1.1(1) \times 10^{-5}$
62	18	2.1×10^{-5}	1.1×10^{-5}	$1.6(1) \times 10^{-7}$
62	19	9.2×10^{-9}	8.9×10^{-6}	$6.2(7) \times 10^{-6}$
62	20	2.0×10^{-8}	1.1×10^{-5}	$2.7(4) \times 10^{-6}$
61	11	2.2×10^{-5}	2.4×10^{-6}	$7.1(1) \times 10^{-5}$
61	12	8.2×10^{-5}	9.1×10^{-5}	$1.0(1) \times 10^{-5}$
61	16	1.0×10^{-4}	5.9×10^{-5}	$6.3(7) \times 10^{-6}$
61	17	4.7×10^{-7}	3.9×10^{-5}	$5.5(8) \times 10^{-5}$
61	18	1.0×10^{-4}	5.0×10^{-5}	$8.7(9) \times 10^{-7}$
61	19	4.3×10^{-8}	4.2×10^{-5}	$3.6(9) \times 10^{-5}$
0	11	2.5×10^{-1}	1.9×10^{-1}	$8.6(9) \times 10^{-2}$

Table 4.2 Transition strengths of weakly bound $(1)0_u^+(J' = 1)$ with weakly and deeply bound $X^1\Sigma_g^+(J = 0)$, given in units of $(ea_0)^2$. Starred v' indicates a heavily Coriolis-mixed excited state. Headers have the same meaning as Table 4.1

$X(v, 0)$	$(1)0_u^+(v', 1)$	Skomorowski	Majewska	Exp. (this work)
62	−4	3.09×10^{-3}	2.77×10^{-3}	$2.57(4) \times 10^{-3}$
61	−4	8.06×10^{-4}	7.36×10^{-4}	$7.0(2) \times 10^{-4}$
61	−5	5.86×10^{-3}	5.06×10^{-3}	$4.30(6) \times 10^{-3}$
60	−6*	7.21×10^{-5}	8.89×10^{-3}	$8.7(4) \times 10^{-3}$
6	−4	2.22×10^{-5}	N/A	$9.9(5) \times 10^{-5}$
4	−5	5.19×10^{-6}	N/A	$2.6(1) \times 10^{-4}$

by a pre-programmed trace (e.g., a positive or negative slope, a Gaussian bell-shaped curve, etc.) on an arbitrary waveform generator fed into a voltage-controlled RF attenuator[5] with sufficiently fast bandwidth. We use separate waveform generators

[5] A double-balanced mixer, such as the Minicircuits ZAD-1+ or ZFM-2-S+, can function as a makeshift current-controlled variable attenuator. The microwave frequency is fed into the RF port. The shaped dc voltage from the arbitrary waveform generator is dropped across a fixed resistor to produce an identically shaped current that is fed into the IF port. The amplitude of the microwave

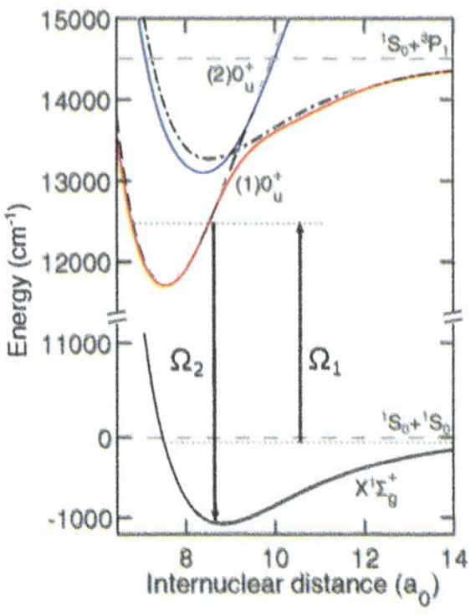

Fig. 4.5 Low-lying states of $(1)0_u^+$ offer an expedient pathway for adiabatic transfer within the ground potential $X^1\Sigma_g^+$. Ω_1 and Ω_2 are the angular Rabi frequencies that characterize the degree of state coupling by the upleg (pump) and downleg (anti-Stokes) lasers, respectively. For visual clarity, horizontal dotted gray lines representing energy levels are not drawn to scale

to independently control each leg and trigger the start of the shaped pulses. Additionally, the microwave signals are sent through separate RF switches (before the respective voltage-controlled attenuators) to completely cut the RF chain and shutter the laser light, depending on the situation.

In the following subsections, we use the 6-state numerical model delineated in Sect. 3.2 to analyze and make sense of our results. We will use the same notation and symbols for the various parameters for consistency. For the spectroscopy of $(1)0_u^+$, only the upleg laser is present. The relevant molecular potentials are shown in Fig. 4.5. For more details on the quantum chemistry of the dimer, see Sect. 2.2.

4.3.1 STIRAP in Free Flight

Figure 4.6a shows a representative time evolution of the number of $X(62, 0)$ molecules during a roundtrip STIRAP at a common detuning of $\Delta = 2\pi \times 30\,\text{MHz}$, and angular Rabi frequencies $|\Omega_1| = 2\pi \times 2.2\,\text{MHz}$ and $|\Omega_2| = 2\pi \times 2.6\,\text{MHz}$. Here, the lattice is tuned to a wavelength of $\lambda_{\text{latt}} = 914.0(1)\,\text{nm}$ and polarized horizontally relative to the MOT quantization axis for optimal in-trap atomic laser cooling on the intercombination line [20]. However, the polarizability difference of $X(62, 0)$ and $X(0, 0)$ at this wavelength introduces lattice-induced thermal decoherence. To overcome this, we perform STIRAP in free flight by shuttering the lattice trap off during the STIRAP pulses. Auxiliary measurements of the molecular cloud size indicate that switching off the trap on timescales of $< 200\,\mu s$ has no

output at the LO port will be proportional to the current at IF. Care should be taken not to exceed the specified damage thresholds. The ZAS-3+, marketed as an attenuator, uses this principle.

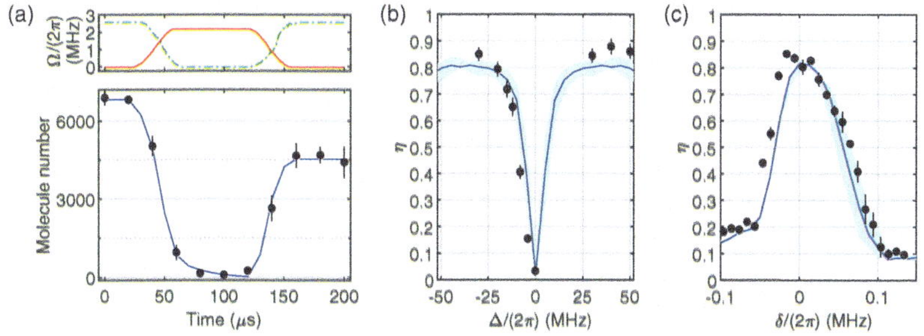

Fig. 4.6 Creation of $^{88}Sr_2$ dimers in the absolute ground state using STIRAP. (**a**) Molecules initially in $X(62,0)$ are transferred to the absolute ground state $X(0,0)$ within $\tau_{tfr} \approx 40\,\mu s$. To detect the transferred molecules, we reverse the order of the STIRAP pulses to recover the initial state. Throughout the roundtrip (200 μs), the optical lattice trap is shuttered off, and the molecules are in free flight. Solid blue line is the model prediction with no free parameters; i.e., we independently measured the time evolution of Ω_1 and Ω_2 (solid red and dashed-dotted green lines respectively), a relative polarization angle of 16(2) degrees, and use $\Gamma_{eff} = 2\pi \times 3.5(5)$ kHz. For this trace, $\Delta/(2\pi) = +30$ MHz. (**b**) One-way transfer efficiency versus the common (one-photon) detuning, Δ. η drops near resonance because of scattering; the dark state is made less robust due to experimental imperfections in controlling laser polarization. (**c**) STIRAP lineshape; η versus the Raman (two-photon) detuning, δ. In both (**b**) and (**c**), the solid blue curve is the model prediction, and the light-blue shaded area covers the range of simulation results, given the uncertainty of the measured parameters

discernable heating effect. Moreover, the first-order Doppler effect is expected to be manageable at the level of $f_{clock}\sqrt{3k_BT/M}/c \sim 3.5(5)$ kHz at temperatures of $T = 8(2)\,\mu K$ for a Raman transition frequency of $f_{clock} \approx 31.825$ THz; M is the molecular mass.

The one-way transfer efficiency is quantified as $\eta = \sqrt{(N_3 - N_2)/N_1}$, where N_1, N_3 are the initial and final molecule numbers respectively, and N_2 is the remaining molecule number after the forward transfer. For our Rabi frequencies, we typically achieve full extinction such that $N_2 = 0$, except in a few extreme cases where Δ is very large, resulting in reduced adiabaticity. Assuming equal forward and reverse transfer efficiencies, we routinely achieve one-way transfer efficiencies of $\eta = 85(3)\%$.

Due to the unresolved Zeeman structure of $(1)0_u^+(11,1)$, a finite relative angle between the polarizations of the Raman (pump and anti-Stokes) lasers undermines dark state formation by coupling the ground states to more than one excited Zeeman sub-level. This leads to increased near-resonant scattering, which diminishes the overall transfer efficiencies. For this dataset, we measure a relative polarization angle of 16(2) degrees between the pump and anti-Stokes, caused by a dichroic mirror combining the lasers with the lattice. We can easily solve this in future work by placing a high-quality polarizer (e.g., Glan-Thompson or Wollaston prism) after this dichroic mirror and immediately before the chamber viewport. Nevertheless, for $J = 0 \leftrightarrow 1$ transitions, we can circumvent this by either lifting the degeneracy of Zeeman sub-levels to realize a three-level Λ-system, or performing STIRAP at Δ

larger than Γ, the linewidth of the $X \rightarrow 0_u^+$ transition. The latter's efficacy is evinced in Fig. 4.6b. Numerical simulations show good agreement by using the measured relative polarization angle and expected Γ_{eff}. Detuned STIRAP also can mitigate other technical imperfections, such as laser phase noise.

When choosing an operational common detuning, it is prudent to survey the molecular structure carefully. For instance, in $^{88}Sr_2$, the rotational splitting of $X(62, 2)$ and $X(62, 0)$ is approximately 70 MHz. Therefore, blue detuning is preferred to avoid accidental perturbation of the excited state by the pump laser should it be tuned close to $X(62, 2) \rightarrow 0_u^+(11, 1)$. Similarly, technical leakage light through the acousto-optic modulators used to modulate the laser intensities can diminish transfer efficiencies should the residual diffraction orders accidentally lie in the vicinity of a $X \rightarrow 0_u^+$ resonance. These technical effects are non-negligible as the laser intensities used in STIRAP are large.

Due to the unequal Rabi frequencies, the detuned-STIRAP lineshape will be asymmetric, where the sharper edge faces the one-photon resonance location [Fig. 4.6c]. The efficiency remains >50% even as the relative Raman detuning is scanned over >100 kHz, and can be widened with greater pump and anti-Stokes laser intensities. Compared to a straightforward Raman π-pulse at similar intensities, STIRAP is robust against small perturbations in laser intensity and laser beam inhomogeneity and is less susceptible to scattering from the intermediate state.

4.3.2 STIRAP in a Magic Wavelength Optical Lattice

We now explore performing STIRAP in a deep optical lattice. Unlike in the previous subsection, here we leave the lattice trap light constantly irradiating throughout the transfer sequence. The trap depth is kept constant at $U_0 = 1009(44)\, E_r$. As illustrated by Fig. 4.7a, our current strategy for engineering magic lattices involves blue-detuning the lattice frequency from a transition connecting the deeply bound ground state with a narrow and isolated rovibronic state in the $(1)1_u$ potential (see Sect. 5.3 for full details). At the magic wavelength, the polarizabilities of the two $J = 0$ vibrational ground states become matched ($\alpha'/\alpha = 1$), resulting in equal trap depths. Doing so removes lattice-induced thermal decoherence and precludes the excitation of breathing modes during the state transfer. We observe the enhancement of STIRAP in a magic wavelength lattice [Fig. 4.7b], achieving one-way efficiencies of over 90%. Figure 4.7c shows the concomitant lattice-induced Stark shift of the peak STIRAP efficiency as the polarizability ratio of the ground states is tuned near the $X \rightarrow 1_u$ transition [Fig. 4.7d].

For $X(0, 0)$, the optimal magic wavelength [17] with the greatest magic detuning ($\Delta_m = 4.494(1)$ GHz) occurs at 1004.7723(1) nm, blue detuned from $X(0, 0) \rightarrow (1)1_u(9, 1)$ with a measured transition strength of $1.335(35) \times 10^{-4}\, (ea_0)^2$ using an all-frequency method (see Sect. 5.3.2). In the Lamb-Dicke regime, the first-order Doppler effect is suppressed, so we set $\Gamma_{\text{eff}} = 0$. In general, the lattice-induced differential light shift on a transition in a non-magic trap becomes compounded for

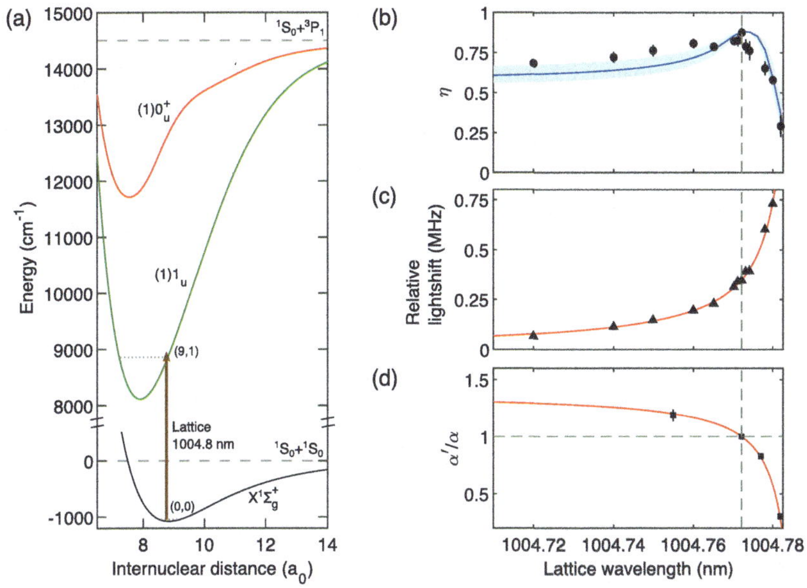

Fig. 4.7 STIRAP transfer in a magic-wavelength optical lattice. (**a**) By tuning the optical lattice near the $(1)1_u(9,1)$, we engineer a magic trap for the pair of states $X(62,0)$ and $X(0,0)$, where their polarizabilities are equal. (**b**) Enhancement of STIRAP efficiency at the magic wavelength: measured (black circles) and modeled (solid blue) with the range of simulation results (shaded light-blue area) for $T = 8(2)\,\mu$K. (**c**) Corresponding lattice-induced light shift on the peak of the STIRAP lineshape, relative to its location in a non-magic lattice tuned >1 THz away. Modifying the ground state polarizabilities shifts the STIRAP lasers out of two-photon resonance. By compensating with either the upleg or downleg laser frequency, the condition $\delta \approx 0$ can be met again. (**d**) Polarizability ratio of $X(0,0)$ to $X(62,0)$ around the magic wavelength, determined from the differential lattice light shifts. In both (**c**) and (**d**), the red solid line is the fit to the data in the form $a/(x - x_0) + b$, where a and b are free parameters, while x_0 is fixed to the independently measured $X(0,0) \to (1)1_u(9,1)$ resonance wavelength

higher trap motional states. To model the transfer efficiencies, we first simulate the ideal STIRAP efficiencies, $\eta(\delta)$, for the measured parameters as a function of δ in the absence of thermal effects. The function $\eta(\delta)$ peaks at approximately $\delta = 0$. For an experimentally set (bare) Raman detuning, δ_{set}, of the pump and anti-Stokes lasers addressing the carrier Raman transition for a molecule in the motional ground state, the thermal probability that a molecule is found in an excited motional trap state maps to a probability density for the molecule to experience a Raman detuning *additionally* shifted by $-\delta'$

$$p(\delta) = p(\delta_{\text{set}} - \delta') = \begin{cases} \frac{1}{2}(B\delta')^2 e^{-B\delta'}, & B\delta' \geq 0 \\ 0, & B\delta' < 0 \end{cases} \tag{4.11}$$

where $B \equiv \frac{h}{k_B T\left(\sqrt{\alpha'/\alpha}-1\right)}$ is a factor that depends on the temperature and polarizability mismatch; essentially the "carrier lineshape" discussed in Sect. 2.4.5. The probability density peaks at $\delta' = \delta'_{\text{max}} = 2/B$. In the experiment, we manually adjust δ_{set} to maximize the STIRAP efficiency (resulting in Fig. 4.7c). It is reasonable to think that this occurs when $\delta_{\text{set}} \simeq \delta'_{\text{max}}$. The overall efficiency thus involves an overlap integral between η and p,

$$\langle\eta\rangle = \frac{\int_{-\infty}^{\infty} d\delta\ \eta(\delta)p(\delta)}{\int_{-\infty}^{\infty} d\delta\ p(\delta)}, \tag{4.12}$$

and the lattice wavelength dependence enters implicitly through α'/α. The solid line in Fig. 4.7b shows the simulation result, reproducing the measurement's salient features fairly well. This thermal model is rather rudimentary, and the astute would notice that the enhancement of the STIRAP efficiency at the magic wavelength exhibits a much sharper feature in the experiment than the model suggests. It is possible that we have inadvertently overlooked additional influences. A Monte-Carlo approach to modeling the same phenomenon would be worthwhile in future work.

4.4 Ultracold Reactive Chemistry with Molecules of Spinless Nuclei

With the creation of large numbers of molecules in $X(0,0)$, we are in a good position to study the basic collisional properties of an ultracold gas of alkaline-earth metal dimers. Homonuclear dimers in the absolute ground state have an exoergic pathway to form stable trimers ($Sr_2 + Sr_2 \rightarrow Sr_3 + Sr$). We, therefore, expect two-body collisions between two $X(0,0)$ molecules to be inelastic and exhibit loss.

Understanding the dynamics of colliding molecules at ultracold temperatures is critical for realizing long-lived bulk samples for various applications in quantum technology (e.g., a molecular clock), and represents an exciting new frontier in the field of cold controlled chemistry [21–23]. So far, every experiment has observed two-body losses of ultracold molecules (e.g., bi-alkali metal molecules, alkaline-earth metal fluorides, etc.), often close to the so-called universal rate. This occurs even when the molecules are prepared in chemically unreactive (e.g., endothermic) quantum states.

Universal loss is a term coined by Idziaszek and Julienne in their seminal paper [24]. They consider the situation when colliding particles are lost with near-unity probability once they come within a characteristic length scale. For collisions between identical molecules in the same state (i.e., $X(0,0)$ in our case), the universal inelastic loss rate via the dominant s-wave channel is independent of the ensemble temperature (T) and given by

$$k_2^{\text{univ}} = 2 \times 2 \times \frac{h}{\mu_4} \times \frac{4\pi}{\Gamma\left(\frac{1}{4}\right)^2} \times \frac{1}{2}\left[\frac{2\mu_4 C_6^{\text{mol}}}{\hbar^2}\right]^{1/4} = \frac{16\pi^2}{\Gamma\left(\frac{1}{4}\right)^2}\left[\frac{2\hbar^2 C_6^{\text{mol}}}{\mu_4^3}\right]^{1/4}, \tag{4.13}$$

where $\mu_4 = (1/(2m_{\text{Sr}}) + 1/(2m_{\text{Sr}}))^{-1}$ is the reduced mass of *two dimers*, $\Gamma(x) = (x-1)!$ is the mathematical gamma function, and C_6^{mol} is the isotropic van der Waals coefficient between *two molecules*.

To calculate C_6^{mol}, our theory collaborators implemented a coupled cluster computation using the explicitly connected representation of the expectation value and polarization propagator [25, 26] and the best approximation XCCSD4 method [27]. The calculations yield $C_6^{\text{mol}} = 15685$ a.u. at the equilibrium distance $R_e = 8.829\,a_0$, which as expected, is approximately four times that of the corresponding coefficient between two atoms. Using Eq. (4.13), we obtain $k_2^{\text{univ}} = 1.22 \times 10^{-10}\,\text{cm}^3\,\text{s}^{-1}$. Note that to convert C_6 from atomic units (a.u.) to units of $\text{cm}^{-1}\text{Å}^6$, the value of C_6 in a.u. will have to be multiplied by $E_{\text{h}}a_0/(hc)$, where $E_{\text{h}} = \hbar^2/(a_0 m_e)$ is the Hartree energy.

4.4.1 Collisions of Ultracold $^{88}Sr_2$ Molecules in the Absolute Ground State

We start by preparing a purified sample of $X(62, 0)$ molecules by wiping away the $J = 2$ molecules from the initial photoassociated mixture. Then, after a forward STIRAP sequence, we hold the $X(0, 0)$ ground state molecules in a non-magic optical lattice ($\lambda_{\text{latt}} = 914.0(1)$ nm) for a variable amount of time, before reversing the STIRAP sequence to recover and detect $X(62, 0)$.

Figure 4.8a shows the decay of the remaining $X(0, 0)$ molecule number per lattice site ($N_{\text{mol/site}}$) over time. We can model the decay with the differential equation[6]

$$\frac{\mathrm{d}n}{\mathrm{d}t} = -k_1 n - k_2 n^2, \tag{4.14}$$

where n is the molecular *density*, which is proportional to $N_{\text{mol/site}}$. The leading order term proportional to n describes the one-body decay, while the term proportional to n^2 describes the two-body decay. It turns out Eq. (4.14) has an analytical solution of the form

[6] In this study, we assume that the temperature of the molecules is unchanged throughout the hold duration. The temperature of the bulk affects the density; i.e., colder samples are denser. More sophisticated analyses typically also model the temperature with a linear time dependence. Doing so leads to a coupled system of differential equations, which has a relatively simple analytical solution [8].

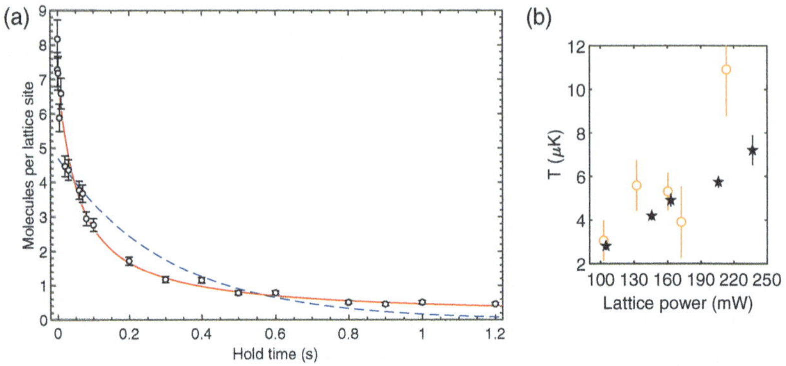

Fig. 4.8 (**a**) Collisional losses of $X^1\Sigma_g^+$ molecules. The number of $X(0, 0)$ molecules (black open circles) decays as they are held in the 1D optical lattice. Fits with the rate equation $dN_{mol/site}/dt = -k_\gamma N^\gamma_{mol/site}$ with $\gamma = 1$ (dashed blue line) and $\gamma = 2$ (solid red line) suggest that the decay is predominantly two-body. The data levels off at around $N_{mol/site} \approx 1/2$, which we ascribe to the equal probability for a given site to be initially populated by an even or odd number of molecules. (**b**) Molecular temperature versus the power in the forward lattice beam, under the experimental conditions of the k_2 loss rate measurements described in this chapter. Black stars, temperature measurement extracted from the thermal broadening of the carrier. Orange circles, auxiliary temperature measurements from the ratio of the axial sideband areas

$$n(t) = \frac{n_0 \exp(-k_1 t)}{1 + (k_2 n_0/k_1)\left[1 - \exp(-k_1 t)\right]}. \tag{4.15}$$

As shown in Fig. 4.8a, the fitted curves strongly indicate two-body collisions to be the dominant loss channel. For $k_2 n \gg k_1$, we may use $\lim_{k_1 \to 0}\left[(1 - e^{-k_1 t})/k_1\right] = t$, which simplifies the expression to

$$n(t) = \frac{n_0}{1 + k_2 n_0 t}. \tag{4.16}$$

From statistical physics, we know that $n = n_{max} \exp\left(-U(y, r)/k_B T\right)$, where $U(y, r)$ is the optical potential and n_{max} is the maximum density at the center of each microtrap. Taking the harmonic approximation [Eq. (2.29)], we write $U(y, r) \approx \frac{1}{2} M \omega_{ax}^2 y^2 + \frac{1}{2} M \omega_{rad}^2 r^2 - U_0$. Therefore, the number of molecules per lattice site is

$$\begin{aligned} N_{mol/site} = n_{max} \int_{-\infty}^{+\infty} dx \int_{-\infty}^{+\infty} dy \int_{-\infty}^{+\infty} dz \exp\left[\frac{-M\omega_{rad}^2 (x^2 + z^2)}{2k_B T}\right] \\ \exp\left[\frac{-M\omega_{ax}^2 y^2}{2k_B T}\right], \\ = n_{max} \left(\frac{2\pi k_B T}{\tilde{\omega}^2 M}\right)^{3/2}, \end{aligned} \tag{4.17}$$

where $\tilde{\omega} \equiv 2\pi(f_{\mathrm{ax}} f_{\mathrm{rad}}^2)^{1/3}$. Similarly, integrating both sides of Eq. (4.14) over volume,

$$\frac{\mathrm{d}}{\mathrm{d}t} N_{\mathrm{mol/site}} = -k_1 N_{\mathrm{mol/site}} - k_2 \left(\frac{\tilde{\omega}^2 M}{4\pi k_B T} \right)^{3/2} N_{\mathrm{mol/site}}^2. \tag{4.18}$$

It is useful to express the rate equation this way because we have direct access to $N_{\mathrm{mol/site}}$ in the experiment; i.e.,

$$N_{\mathrm{mol/site}} = N_{\mathrm{mol}} / N_{\mathrm{sites}}, \tag{4.19}$$

where N_{mol} is given by Eq. (2.3) and N_{sites} is the number of filled sites.[7]

4.4.2 *Two-Body Loss Rate*

Since the molecules are tightly trapped in the Lamb-Dicke and resolved sideband regimes, we can spectroscopically access the axial and radial trap frequencies of $X(62, 0)$ by probing the shallow-to-shallow Raman transition $X(62, 0) \rightarrow X(61, 0)$; see Sect. 2.4.4. Importantly, we must further scale the *square* of the trap frequencies of $X(62, 0)$ by the polarizability ratio of $X(0, 0)$ to $X(62, 0)$, α'/α, in order to properly obtain $\tilde{\omega}^2$ for $X(0, 0)$ as it appears in Eq. (4.18).

From measurements of differential light shift [Eq. (2.47)] for the shallow-to-deep Raman transition $X(62, 0) \rightarrow X(0, 0)$, we find $\alpha'/\alpha = 1.5176(59)$ at $\lambda_{\mathrm{latt}} = 914.0(1)$ nm. We use the same data to determine the temperature of the molecular ensemble via carrier thermometry; see Sect. 2.4.5. Under the experimental conditions during this particular measurement, we compare the temperatures extracted using carrier thermometry to the conventional method of taking the ratio of the integrated area under the sidebands and find good agreement, as shown in Fig. 4.8b.

The measured two-body loss rate coefficient k_2 for $X(0, 0)$ molecules is $0.97(28) \times 10^{-10}\,\mathrm{cm^3\,s^{-1}}$. We additionally observe that k_2 is insensitive to the presence of $X(62, 2)$ rotationally excited near-threshold molecules in the trap. The closeness of our measured k_2 to the universal loss rate (k_2^{univ}) implies that alkaline-earth metal dimers in the absolute ground state are lost with near unity probability following a collision at short range.

[7] Under normal conditions, the number of filled sites is roughly ~550, with no discernable change over the range of lattice laser powers used for the experiments in this thesis. We determine this number by imaging the photofragments of recovered molecules from the "side" of the lattice. For simplicity, we will assume equal occupancy across sites. Higher resolution imaging should allow the characterization of density gradients along the lattice axis in future work if required.

4.4.3 Losses in Other Ultracold Molecules

Despite fervent theoretical and experimental efforts, a complete picture of how ultracold molecules are lost due to collisions remains equivocal. For comprehensive and contemporary reviews, see Refs. [23, 28]. The prevailing sentiment is that the crux of the matter lies in the formation of intermediate four-body (tetramer) complexes when two diatomic molecules collide [29, 30]. Several mechanisms have been put forward for how complex formation leads to the observed losses in experiments.

In the "sticky collisions" hypothesis, a subsequent collision of these delicate complexes with the surrounding dimers produces sufficient kinetic energy to eject all three molecules out of the trap. Crucially, the sticking time sets a finite window for an ejection collision. In principle, calculations of the sticking time via Rice-Ramsperger-Kassel-Marcus (RRKM) theory should enable quantitative tests with experiments [31].

A common feature among the experiments is that a near-IR wavelength optical trap levitates the molecules against gravity. Another hypothesis proposes that the intermediate complexes interact with the trapping laser to undergo photofragmentation or photoexcitation into electronically excited states [32]. For the typical intensities used for optical trapping, it was estimated that the absorption of laser light occurs much faster than the rate at which the collision complex decays back into two ground state dimers, leading to effective two-body loss. This argumentation has been gaining traction. The interaction of the complex with trap light has been experimentally confirmed in KRb and RbCs molecules [23, 33, 34]. These recent experiments chopped the trap laser intensity to create durations where the molecules do not experience the trapping light. However, parallel experiments with similar molecular species [35, 36] did not find evidence of such photoinduced loss, suggesting that either theoretical calculations have severely underestimated the complex lifetimes, or that additional loss mechanisms may be present. One emerging line of thought is that non-conservation of nuclear spin may considerably increase the density of tetramer states, resulting in significantly longer complex lifetimes [37]. To help elucidate the issue on nuclear spins, Jachymski et al. [37] has proposed experiments with mixtures of alkaline-earth metal molecules and atoms, making use of the fact that $^{88}Sr_2$ has no net nuclear spin, while the isotopologue $^{87}Sr_2$ possesses hyperfine structure ($9/2 \oplus 9/2 = 0, 1, \ldots, 9$).

Similarly, among the various ultracold molecular gases created so far, the collisional chemistry of alkaline-earth metal dimers (without nuclear spin) appears the simplest to model. Not only can the total angular momentum of the molecule be zero, but the electronic ground rovibrational states are much less sensitive to external electric and magnetic fields than bi-alkali metal molecules, ensuring angular momentum conservation during a collision (an assumption in the theoretical estimate of the sticking times so far). More detailed studies with Sr_2 should help verify the central principles of RRKM theory when applied to ultracold molecular gases.

Whatever the cause, microwave shielding of dipolar molecules has been demonstrated to be a viable method to circumvent inelastic collisional losses [38, 39]. Conceptually, an off-resonant external microwave field induces a rotating dipole in the molecule, forming a potential barrier between two molecules at short range. In a remarkable advancement for the field, engineered elastic collisional rates through microwave-shielding have led to efficient evaporative cooling of a molecular gas to quantum degeneracy [39]. For homonuclear Sr_2 molecules, however, these methods are not directly applicable; only even J are allowed in $X^1\Sigma_g^+$ for the bosonic Sr isotope (see Sect. 2.2), precluding *dipole* microwave dressing using excited rotational states. Moreover, the field dressing inherently introduces light shifts that would be an additional source of systematic error in precision molecular clock applications.

References

1. Bergmann, K., Vitanov, N.V., Shore, B.W.: Perspective: stimulated Raman adiabatic passage: the status after 25 years. J. Chem. Phys. **142**, 170901 (2015)
2. Bergmann, K., et al.: Roadmap on STIRAP applications. J. Phys. B: At. Mol. Opt. Phys. **52**, 202001 (2019)
3. Bergmann, K., Theuer, H., Shore, B.: Coherent population transfer among quantum states of atoms and molecules. Rev. Mod. Phys. **70**, 1003 (1998)
4. Vitanov, N.V., Rangelov, A.A., Shore, B.W., Bergmann, K.: Stimulated Raman adiabatic passage in physics, chemistry, and beyond. Rev. Mod. Phys. **89**, 015006 (2017)
5. Wenz, K.: Nuclear Schiff moment search in thallium fluoride molecular beam: rotational cooling. PhD thesis (Columbia University, New York, 2021)
6. Bennink, R.S.: Frequency conversion of optical signals using coherently prepared media. PhD thesis (University of Rochester, Rochester, 2004)
7. Yatsenko, L., Romanenko, V., Shore, B., Bergmann, K.: Stimulated Raman adiabatic passage with partially coherent laser fields. Phys. Rev. A **65**, 043409 (2002)
8. Bause, R.: Understanding and controlling the collisions of ultracold polar molecules. PhD thesis (Ludwig Maximilian University of Munich, Munich, 2022)
9. Yatsenko, L., Shore, B., Bergmann, K.: Detrimental consequences of small rapid laser fluctuations on stimulated Raman adiabatic passage. Phys. Rev. A **89**, 013831 (2014)
10. Seeßelberg, F.: Interacting gases of ultracold polar molecules. PhD thesis (Ludwig Maximilian University of Munich, Munich, 2019)
11. Seeßelberg, F., et al.: Modeling the adiabatic creation of ultracold polar $^{23}Na^{40}K$ molecules. Phys. Rev. A **97**, 013405 (2018)
12. Bause, R., et al.: Efficient conversion of closed-channel-dominated Feshbach molecules of $^{23}Na^{40}K$ to their absolute ground state. Phys. Rev. A **104**, 043321 (2021)
13. Zare, R.N., Harter, W.G.: Angular Momentum: Understanding Spatial Aspects in Chemistry and Physics. Wiley-Interscience, Hoboken (1988)
14. Hansson, A., Watson, J.K.: A comment on Hönl-London factors. J. Mol. Spectrosc. **233**, 169–173 (2005)
15. Skomorowski, W., Moszynski, R., Koch, C.P.: Formation of deeply bound ultracold Sr_2 molecules by photoassociation near the $^1S + {}^3P_1$ intercombination line. Phys. Rev. A **85**, 043414 (2012)
16. Skomorowski, W.: Production of ultracold molecules by photoassociation spectroscopy and sympathetic cooling: a theoretical study PhD thesis (University of Warsaw, Warsaw, 2013)

17. Leung, K.H., et al.: Transition strength measurements to guide magic wavelength selection in optically trapped molecules. Phys. Rev. Lett. **125**, 153001 (2020)
18. Skomorowski, W., Pawłowski, F., Koch, C.P., Moszynski, R.: Rovibrational dynamics of the strontium molecule in the $A^1\Sigma_u^+$, $c^3\Pi_u$, and $a^3\Sigma_u^+$ manifold from state-of-the-art ab initio calculations. J. Chem. Phys. **136**, 194306 (2012)
19. Kondov, S.S., et al.: Molecular lattice clock with long vibrational coherence. Nat. Phys. **15**, 1118–1122 (2019)
20. Ido, T., Katori, H.: Recoil-free spectroscopy of neutral Sr atoms in the Lamb-Dicke regime. Phys. Rev. Lett. **91**, 053001 (2003)
21. Heazlewood, B.R., Softley, T.P.: Towards chemistry at absolute zero. Nat. Rev. Chem. **5**, 125–140 (2021)
22. Krems, R.V.: Cold controlled chemistry. Phys. Chem. Chem. Phys. **10**, 4079–4092 (2008)
23. Liu, Y., Ni, K.-K.: Bimolecular chemistry in the ultracold regime. Annu. Rev. Phys. Chem. **73**, 73–96 (2022)
24. Idziaszek, Z., Julienne, P.S.: Universal rate constants for reactive collisions of ultracold molecules. Phys. Rev. Lett. **104**, 113202 (2010)
25. Jeziorski, B., Moszynski, R.: Explicitly connected expansion for the average value of an observable in the coupled-cluster theory. Int. J. Quantum Chem. **48**, 161–183 (1993)
26. Moszynski, R., Żuchowski, P.S., Jeziorski, B.: Time-independent coupled-cluster theory of the polarization propagator. Collect. Czechoslov. Chem. Commun. **70**, 1109–1132 (2005)
27. Korona, T., Przybytek, M., Jeziorski, B.: Time-independent coupled cluster theory of the polarization propagator. Implementation and application of the singles and doubles model to dynamic polarizabilities and van der Waals constants. Mol. Phys. **104**, 2303–2316 (2006)
28. Bause, R., Christianen, A., Schindewolf, A., Bloch, I., Luo, X.-Y.: Ultracold sticky collisions: theoretical and experimental status. J. Phys. Chem. A **127**, 729–741 (2023)
29. Mayle, M., Quéméner, G., Ruzic, B.P., Bohn, J.L.: Scattering of ultracold molecules in the highly resonant regime. Phys. Rev. A **87**, 012709 (2013)
30. Hu, M.-G., et al.: Direct observation of bimolecular reactions of ultracold KRb molecules. Science **366**, 1111–1115 (2019)
31. Gregory, P.D., et al.: Sticky collisions of ultracold RbCs molecules. Nat. Commun. **10**, 1–7 (2019)
32. Christianen, A., Zwierlein, M.W., Groenenboom, G.C., Karman, T.: Photoinduced two-body loss of ultracold molecules. Phys. Rev. Lett. **123**, 123402 (2019)
33. Liu, Y., et al.: Photo-excitation of long-lived transient intermediates in ultracold reactions. Nat. Phys. **16**, 1132–1136 (2020)
34. Gregory, P.D., Blackmore, J.A., Bromley, S.L., Cornish, S.L.: Loss of ultracold $^{87}Rb^{133}Cs$ molecules via optical excitation of long-lived two-body collision complexes. Phys. Rev. Lett. **124**, 163402 (2020)
35. Bause, R., et al.: Collisions of ultracold molecules in bright and dark optical dipole traps. Phys. Rev. Res. **3**, 033013 (2021)
36. Gersema, P., et al.: Probing photoinduced two-body loss of ultracold nonreactive Bosonic $^{23}Na^{87}Rb$ and $^{23}Na^{39}K$ molecules. Phys. Rev. Lett. **127**, 163401 (2021)
37. Jachymski, K., Gronowski, M., Tomza, M.: Collisional losses of ultracold molecules due to intermediate complex formation. Phys. Rev. A **106**, L041301 (2022)
38. Anderegg, L., et al.: Observation of microwave shielding of ultracold molecules. Science **373**, 779–782 (2021)
39. Schindewolf, A., et al.: Evaporation of microwave-shielded polar molecules to quantum degeneracy. Nature **607**, 677–681 (2022)

Chapter 5
Terahertz Vibrational Molecular Clock

5.1 Polarizability and Transition Strengths

5.1.1 Preliminary Comments

Generally speaking, under the application of an electric field, a particle's charge distribution may be expanded in terms of its electric multipole moments. If the particle is neutral (i.e., not a monopole), the leading non-trivial term is the *dipole* moment, which itself can be expanded [1] in increasing orders of the electric field $\mathcal{E}_j(t) = \frac{\mathcal{E}_0}{2}\left(\epsilon_j\, e^{-i\omega t} + \epsilon_j^*\, e^{+i\omega t}\right)$,

$$d_i(t) = d_{\text{perm}} + \alpha_{ij}\mathcal{E}_j(t) + \frac{1}{2!}\beta_{(1),ijk}\mathcal{E}_j(t)\mathcal{E}_k(t) + \frac{1}{3!}\beta_{(2),ijkl}\mathcal{E}_j(t)\mathcal{E}_k(t)\mathcal{E}_l(t) + \ldots. \tag{5.1}$$

As with the previous chapters, ϵ_j is the unit vector pointing along the polarization of the electric field, and $\mathcal{E}_0$ is the field amplitude. Here, I adopt the Einstein summation convention, where repeated indices are implied to be summed over. While orthogonal, these Cartesian indices do not necessarily mean $\hat{\mathbf{X}}$, $\hat{\mathbf{Y}}$, or $\hat{\mathbf{Z}}$. The first term d_{perm} is the permanent dipole moment. The remaining terms characterize the *induced* dipole moment, starting with the lowest-order term containing the second-rank tensor α_{ij} called the **polarizability**. The higher order corrections (e.g., $\beta_{(1),ijk}$, $\beta_{(2),ijkl}$, ...) are collectively referred to as hyperpolarizabilities. For a homonuclear dimer in the electronic ground state, $\mathbf{d}$ is an odd function in $\mathcal{E}$, so $d_{\text{perm}} = 0$ and $\beta_{(n),ijk} = 0$ for odd (n). From hereon, we will only keep the **second hyperpolarizability** $\beta_{(2),ijkl} \equiv \gamma_{ijkl}$ and drop its (2) subscript for brevity.

The charge redistribution in the particle must be accompanied by a change in its internal energy,

K. H. Leung, *The Strontium Molecular Lattice Clock*, Springer Theses,
https://doi.org/10.1007/978-3-031-47647-1_5

$$
\begin{aligned}
U(t) &= -\int_0^{\mathcal{E}} [d_i']^* \, \mathrm{d}\mathcal{E}_i' \\
&= -\alpha_{ij}\epsilon_i^*\epsilon_j \int_0^{\mathcal{E}} \mathcal{E}' \mathrm{d}\mathcal{E}' - \frac{1}{3!}\gamma_{ijkl}\epsilon_i^*\epsilon_j^*\epsilon_k^*\epsilon_l \int_0^{\mathcal{E}} \mathcal{E}'^3 \mathrm{d}\mathcal{E}' \\
&= -\frac{\mathcal{E}_0^2\cos^2(\omega t)}{2!}\alpha_{ij}\epsilon_i^*\epsilon_j - \frac{\mathcal{E}_0^4\cos^4(\omega t)}{4!}\gamma_{ijkl}\epsilon_i^*\epsilon_j^*\epsilon_k^*\epsilon_l.
\end{aligned}
$$

In the above expression for the internal energy, it may be more rigorous to add the complex conjugate and take the average [2], which we have left out for brevity; i.e.,

$$
U(t) = -\frac{1}{2}\left(\int [d_i']^* \, \mathrm{d}\mathcal{E}_i' + \int d_i' \, \mathrm{d}\left[\mathcal{E}_i'\right]^*\right).
$$

Note that the numerical prefactors in Eq. (5.1) are chosen so that [3, 4]

$$
\alpha_{ij} = -\left.\frac{\partial^2 U(t)}{\partial\mathcal{E}_i(t)\partial\mathcal{E}_j(t)}\right|_{\mathcal{E}_0=0}, \quad \gamma_{ijkl} = -\left.\frac{\partial^4 U(t)}{\partial\mathcal{E}_i(t)\partial\mathcal{E}_j(t)\partial\mathcal{E}_k(t)\partial\mathcal{E}_l(t)}\right|_{\mathcal{E}_0=0}, \quad \text{etc.}
$$

As before (see Sect. 2.4.1), we take the time average over 1-cycle to obtain the energy shift [5]

$$
U = \langle U(t)\rangle_{\text{time averaged}} = -\frac{\mathcal{E}_0^2}{4}\alpha_{ij}\epsilon_i^*\epsilon_j - \frac{\mathcal{E}_0^4}{64}\gamma_{ijkl}\epsilon_i^*\epsilon_j^*\epsilon_k^*\epsilon_l, \tag{5.2}
$$

which is a more sophisticated way of writing Eq. (2.4) by including the laser polarization dependence as well as the hyperpolarizability.

So far, we have written α_{ij}, γ_{ijkl}, ϵ_i as Cartesian tensors. It is also possible to write them as spherical tensors. As far as a physicist is concerned, we can define a spherical tensor operator of rank L to be the set of objects $Q_K^{(L)}$ with $2L+1$ components ranging between $-L \le K \le L$ that transform under rotation in exactly the same way as an angular momentum eigenket $|J = L, M = K\rangle$. Since spherical harmonics are simply angular momentum eigenkets projected onto the polar and azimuth angles, $Y_M^J(\theta, \phi) \equiv \langle\theta, \phi|J, M\rangle$, one arrives at the useful fact that $Q_K^{(L)}$ is proportional to $Y_K^L(\theta, \phi)$.

In addition, we know from elementary quantum mechanics that the coupling of two angular momentum states $|J_1, M_1\rangle$ and $|J_2, M_2\rangle$ into a resultant third $|J_3, M_3\rangle$ involves a summation weighted by Clebsch-Gordon coefficients (essentially change-of-basis amplitudes),

$$
|J_3, M_3\rangle = \sum_{M_1, M_2} \langle J_1, M_1; J_2, M_2|J_3, M_3\rangle \, |J_1, M_1; J_2, M_2\rangle .
$$

Therefore, by definition, the same expression must also be true for spherical tensor operators,[1]

$$Q_{K_3}^{(L_3)} = \sum_{K_1,K_2} \langle L_1, K_1; L_2, K_2 | L_3, K_3 \rangle \, Q_{K_1}^{(L_1)} \, Q_{K_2}^{(L_2)}. \tag{5.3}$$

Equation (5.3) tells us how to 'make' a spherical tensor from the product of another two. One consequence is the general statement that scalar (rank 0) operators can be constructed by contracting two tensor operators of the same rank; e.g., energy U is a scalar.

5.1.2 Scalar, Vector, and Tensor Polarizabilities

To see how the polarizability tensor α_{ij} has 'structure', let us rewrite Eq. (5.2) in spherical tensor form (ignoring the hyperpolarizability for now),

$$U = -\frac{\mathcal{E}_0^2}{4} \sum_{L=0}^{2} \sum_{K=-L}^{+L} (-1)^K \alpha_K^{(L)} \xi_{-K}^{(L)},$$

where $\alpha_K^{(L)}$ is the K component of a rank L tensor. The *polarization* tensor is defined as [2]

$$\xi_K^{(L)} \equiv \sqrt{2L+1} \sum_{\mu=0,\pm 1} (-1)^{\mu} \epsilon_\mu \epsilon_{\mu-K}^* \begin{pmatrix} 1 & L & 1 \\ \mu & -K & K-\mu \end{pmatrix}, \tag{5.4}$$

which can be obtained (up to conventional numerical factors) by applying Eq. (5.3) to the case of a dyadic $(\epsilon_i^* \epsilon_j)$. Note that $(\epsilon^*)_\mu = (-1)^\mu (\epsilon_{-\mu})^*$. Often the quantum state under consideration has a well-defined angular momentum and parity (e.g., $|J, M\rangle$). We may invoke the correspondence principle to write $\alpha_K^{(L)}$ as the matrix element $\langle J, M | \mathcal{A}_K^{(L)} | J, M \rangle$, where $\mathcal{A}_K^{(L)}$ is the polarizability *operator*. The Wigner-Eckart theorem tells us that

$$\langle J, M | \mathcal{A}_K^{(L)} | J, M \rangle \propto \begin{pmatrix} J & L & J \\ -M & K & M \end{pmatrix} \neq 0,$$

if and only if $K = 0$, by virtue of the triangle condition. Hence, we may collapse the formula for the energy shift to contain only three $K = 0$ terms

[1] Some authors omit the sum over K_2 explicitly since the triangle condition restricts non-vanishing terms to have certain values of K_2 related to K_1.

$$U = -\frac{\mathcal{E}_0^2}{4} \sum_{L=0}^{2} (-1)^L \langle J, M | \mathcal{A}_0^{(L)} | J, M \rangle \, \xi_0^{(L)}. \tag{5.5}$$

The so-called scalar, vector, and tensor shifts are related to the $L = 0, 1, 2$ terms, respectively.

We may further factor out the J and M dependence of the various terms using the Wigner-Eckart theorem [6]; i.e., the matrix element of a spherical tensor operator is proportional to that of the angular momentum spherical tensor operator of the same rank and component, $J_K^{(L)}$. In other words, $\langle J, M | \mathcal{A}_K^{(L)} | J, M \rangle = c_L \langle J, M | J_K^{(L)} | J, M \rangle$, where $c_{L=0,1,2}$ are constants of proportionality and may be positive or negative. It is customary to collect the prefactors and redefine them as the real numbers $\alpha_{\text{sc}}, \alpha_{\text{vec}}, \alpha_{\text{tns}}$ where the subscripts denote 'scalar', 'vector', and 'tensor' respectively.

A detailed treatment can be found in Ref. [2]. In the perturbative regime where the total energy shift U due to the laser is much smaller than the energy separation of the molecular states (so that J and M remain good quantum numbers),

$$U = -\frac{\mathcal{E}_0^2}{4} \left(\alpha_{\text{sc}} - \alpha_{\text{vec}} \frac{M}{2J} i(\boldsymbol{\epsilon}^* \times \boldsymbol{\epsilon}) \cdot \hat{\mathbf{q}} + \alpha_{\text{tns}} \frac{3M^2 - J(J+1)}{J(2J-1)} \frac{3|\boldsymbol{\epsilon} \cdot \hat{\mathbf{q}}|^2 - 1}{2} \right). \tag{5.6}$$

For the state $|i\rangle$ with zero nuclear spin ($I = 0$), the explicit expressions for the scalar, vector, and tensor polarizabilities in terms of the reduced matrix elements $\langle f || d^{(1)} || i \rangle$ are

$$\alpha_{\text{sc}} = \frac{1}{\hbar} \frac{1}{3(2J+1)} \sum_f |\langle f || d^{(1)} || i \rangle|^2 \frac{2\omega_{fi}}{\omega_{fi}^2 - \omega^2}, \tag{5.7}$$

$$\alpha_{\text{vec}} = -\frac{1}{\hbar} \sqrt{\frac{6J}{(2J+1)(J+1)}} \sum_f (-1)^{J+J'} \begin{Bmatrix} 1 & 1 & 1 \\ J & J' & J \end{Bmatrix}$$
$$|\langle f || d^{(1)} || i \rangle|^2 \frac{2\omega}{\omega_{fi}^2 - \omega^2}, \tag{5.8}$$

$$\alpha_{\text{tns}} = \frac{1}{\hbar} \sqrt{\frac{10J(2J-1)}{3(2J+1)(J+1)(2J+3)}} \sum_f (-1)^{J+J'} \begin{Bmatrix} 1 & 2 & 1 \\ J & J' & J \end{Bmatrix}$$
$$|\langle f || d^{(1)} || i \rangle|^2 \frac{2\omega_{fi}}{\omega_{fi}^2 - \omega^2}, \tag{5.9}$$

where $|i\rangle$, $|f\rangle$ are shorthand labels for $|n_i, v_i; J_i, M_i, \Omega_i\rangle$, $|n_f, v_f; J_f, M_f, \Omega_f\rangle$; i.e., the full molecular state. In the formulas above, we relabeled $J_i \to J$, $M_i \to M$, $J_f \to J'$ and $M_f \to M'$ for notation consistency within this subsection. The array with curly brackets is the Wigner 6-j symbol. ω_{fi} is the corresponding transition angular frequency,[2] and ω is the angular frequency of the electromagnetic field. We point out that the angular frequency that appears in the numerator of Eq. (5.8) differs from Eqs. (5.7) and (5.9); i.e., it is ω and not ω_{fi}. The dc (static) polarizability is recovered for $\omega = 0$. While this sum-over-states approach is popular and intuitive, other methods to numerically calculate molecular polarizability also exist [1, 7]. In particular, often it is necessary to add a small contribution, α_{core}, due to the core electrons to the scalar polarizability [8].

For the work in this thesis it is appropriate to write the polarization vector [Eq. (3.8)] of the laser propagating along $\hat{\mathbf{Y}}$ as

$$\boldsymbol{\epsilon} = \hat{\mathbf{Z}} \cos\theta + \hat{\mathbf{X}} e^{i\phi} \sin\theta,$$

where θ is the polarization angle[3] with respect to the lab frame quantization axis $\hat{\mathbf{q}} = \hat{\mathbf{Z}}$ (i.e., $\boldsymbol{\epsilon} \cdot \hat{\mathbf{q}} = \cos\theta$), and ϕ is the Bloch sphere azimuth angle that parametrizes the degree of circular polarization. Our trap laser propagates perpendicular relative to the quantization axis in this work, so the polarization dependence of the vector light shift does not have a clean expression[4] purely in terms of ϕ. As Eq. (5.6) involves lab-fixed frame transition matrix elements, it gives the total light shift due to the polarizability of the particle in the lab-fixed frame.[5]

The appearance of the Legendre polynomial $P_2(\cos\theta) = (3\cos^2\theta - 1)/2$ in the tensor light shift is not surprising given its rank 2 nature. At the "magic angle" $\theta = \cos^{-1}\left(1/\sqrt{3}\right) \approx 54.7°$, the tensor light shift can be suppressed to zero (see e.g., Ref. [10] for the subradiant $(1)1_g$ states). For $J = 1/2$, the tensor light shift is zero. For $J = 0$, both the vector and tensor light shifts are zero. For $M = 0$, or linearly polarized light, or when the projection of $(\boldsymbol{\epsilon}^* \times \boldsymbol{\epsilon})$ along the quantization axis ($\hat{\mathbf{q}}$) is zero, the vector light shift is zero. α_{sc} is also called the isotropic polarizability, and the trace of the polarizability gives the rank 0 part, since $\operatorname{tr}\alpha_0^{(L)} = \sum_M \langle J, M|\alpha_0^{(L)}|J, M\rangle \propto \alpha_0^{(0)}$ as terms with $L \neq 0$ vanish by the Wigner-Eckart theorem. This is reflected in the fact that, individually, the vector and tensor light shifts sum to zero over all possible M states,

---

[2] Note that ω_{fi} is positive (negative) if $|f\rangle$ is energetically higher (lower) than $|i\rangle$.

[3] Here, the laser propagates *perpendicular* to the quantization axis, allowing us to make an intuitive correspondence between the geometric polarization angle relative to quantization axis ($= \theta$) and the Bloch sphere polar angle ($= 2\theta$).

[4] For a laser propagating *parallel* to the $\hat{\mathbf{Z}}$ quantization axis, the Bloch vector points toward the pole ($2\theta = \pi$). Here, ϕ loses its meaning and it is more appropriate write $\boldsymbol{\epsilon} = \hat{\mathbf{X}} \cos\gamma + \hat{\mathbf{Y}}\, i \sin\gamma$, where γ is the so-called degree of ellipticity. The cross product $-i(\boldsymbol{\epsilon}^* \times \boldsymbol{\epsilon}) \cdot \hat{\mathbf{Z}}$ would evaluate to $\sin(2\gamma)$.

[5] For molecules, it is important to distinguish between the lab-fixed frame and the molecule-fixed frame. In particular, for ${}^1\Sigma_g^+$ molecules, the analogous molecular frame 'scalar', 'vector', and 'tensor' polarizabilities are $\tilde{\alpha}_{\text{sc}} = \frac{1}{3}(\tilde{\alpha}_{\parallel} + 2\tilde{\alpha}_{\perp})$, $\tilde{\alpha}_{\text{vec}} = 0$, $\tilde{\alpha}_{\text{tns}} = \frac{2}{3}(\tilde{\alpha}_{\parallel} - \tilde{\alpha}_{\perp})$ respectively. Here, $\tilde{\alpha}_{\parallel}$ and $\tilde{\alpha}_{\perp}$ are the molecular frame polarizabilities involving parallel Σ-Σ and perpendicular Σ-Π transitions respectively [1, 9].

$$\sum_{M=-J}^{+J}\sum_{L=1}^{2}(-1)^L\,\langle J,M|\mathcal{A}_0^{(L)}|J,M\rangle\,\xi_0^{(L)}$$
$$= \text{const.}\times\sum_{M=-J}^{+J}M+\text{const.}\times\sum_{M=-J}^{+J}[3M^2-J(J+1)]$$
$$=0.$$

The first term is evidently zero. For the second term, we used the math result for the square pyramidal numbers: $\sum_{M=-J}^{+J}M^2=J(J+1)(2J+1)/3$.

The essence of the above discussion that pertains to this thesis is that we expect the polarizabilities of $X^1\Sigma_g^+(J=0)$ clock states to be scalar; i.e., the light shift on these states arising from a finite **polarizability** is **independent** of laser polarization. However, in general, the light shift arising from the **hyperpolarizability** is a function of the laser's **degree of circular polarization**, even for $J=0$ states [3, 4]. This can be shown from Eq. (5.2), but a simple physical argument is sufficient to illustrate this. Consider a consecutive two-photon transition (Ξ-configuration) from $J=0$ to $J''=0$ via $J'=1$. If the laser is linearly polarized, the hyperpolarizability light shift of the initial $J=0$ state exhibits a dispersive resonance at the two-photon resonance. However, this two-photon transition is clearly forbidden if the laser is circularly polarized (the second leg could not address the final state), and the light shift is unperturbed by the two-photon resonance. Exactly such a two-photon transition between scalar states is known to contribute to the hyperpolarizability shifts in the Yb atomic optical lattice clock at its 759 nm magic wavelength [11, 12].

5.1.3 Clock State Polarizability Formula

The **electric dipole** ($E1$) polarizabilities of $J=0$ clock states in the $X^1\Sigma_g^+$ potential were calculated by our theory collaborators [13]. Using Eq. (5.7), the purely scalar polarizability for a state with vibrational quantum number υ is

$$\alpha_\upsilon=\frac{1}{\hbar}\sum_f S_{fi}\left(\frac{1}{\omega_{fi}-\omega}+\frac{1}{\omega_{fi}+\omega}\right). \tag{5.10}$$

As in Eqs. (4.4) and (4.5), the transition strengths are

$$S_{fi}\equiv\left|\left\langle n_f,\upsilon_f;J_f,M_f,\Omega_f\left|d^{(1)}_{M_f-M_i}\right|n_i,\upsilon_i;J_i,M_i,\Omega_i\right\rangle\right|^2. \tag{5.11}$$

Here, $J_i=0$, $M_i=0$, $\Omega_i=0$, $n_i=X^1\Sigma_g^+$, and $J_f=1$ such that

$$S_{fi} = \begin{pmatrix} 1 & 1 & 0 \\ 0 & 0 & 0 \end{pmatrix}^2 |\langle f || d^{(1)} || i \rangle|^2 = \frac{1}{3} |\langle f || d^{(1)} || i \rangle|^2.$$

The sum in Eq. (5.10) is mainly composed of transitions from $n_i = X^1\Sigma_g^+$ to singlet ungerade excited potentials (n_f). Bound-to-continuum transitions are calculated by discretizing the continuum. The n_f channels include the ${}^1\Sigma_u^+$ potentials correlating to ${}^1S + {}^1P$ and ${}^1S + {}^1D$, as well as the ${}^1\Pi_u$ potentials correlating to ${}^1S + {}^1P$, ${}^1S + {}^1D$, ${}^3P + {}^3P$ and ${}^3P + {}^3D$. The $(1)^1\Sigma_u^+$ potential is taken from the *ab initio* calculation in Ref. [14], while the doubly-excited $(3)^1\Pi_u$ and $(4)^1\Pi_u$ potentials were calculated using the multireference configuration interaction method (MRCI) with the MOLPRO package [15]. The remaining potentials (including $X^1\Sigma_g^+$) are empirical [16, 17]. We omit spin-orbit and non-adiabatic couplings between the potentials. The convergence of our results is not changed by the inclusion of further high-lying potentials [13].

Laser wavelengths in the range 800–1200 nm can drive transitions from $X^1\Sigma_g^+$ to the short-range part of $(1)1_u$. These singlet-triplet transitions are relatively weaker than singlet-singlet ones, but become important if the laser is tuned near a resonance; e.g., in the case of a magic wavelength optical trap. To properly account for these situations, we additionally include the Morse/Long-range potential of $(1)1_u$ from Ref. [18] in the polarizability sum.

Classically, the polarizability is proportional to the particle's spatial volume (in fact, it has the same dimensions as volume in c.g.s units). For this reason, the polarizability of a weakly bound homonuclear diatomic molecule in $X^1\Sigma_g^+$ will be approximately *twice* that of a single constituent ${}^1S_0$ atom; i.e., $\alpha_{62} \simeq 2 \times \alpha_{\mathrm{Sr}}$. This allows us to benchmark the absolute accuracy of our polarizability calculations with highly accurate values reported in Sr atomic clock literature. We estimate that the polarizability of $X(62, 0)$ is 2×282 a.u. at 813.4 nm and 2×256 a.u. at 914.0 nm. These are in reasonable agreement ($\sim 2\%$) with the values for α_{Sr} obtained experimentally (288 a.u. at 813.4 nm [19]) and theoretically (286.0(3) a.u. at 813.4 nm [20] and 261.2(3) a.u. at 914.0 nm [20]).

Additionally, we can benchmark the relative accuracy of our calculations with ac polarizability ratios α_v/α_{62} for various v, over a range of lattice wavelengths (e.g., $v = 0$ in Fig. 5.1). We do this using a frequency-only method [Eq. (2.47)], finding consistency within $< 20\%$.

As a sanity check, let us make contact with simpler-looking formulas from earlier in this thesis. Suppose we have a two-level system coupled by a laser. Since we are dealing with optical frequencies, $1/(\omega_{fi} + \omega) \ll 1/(\omega_{fi} - \omega) \equiv -1/\Delta_l$ and Eq. (5.10) reduces to just a single term $\alpha = -S/(\hbar\Delta_l)$. Using Eqs. (4.3) and (2.4), the angular frequency shift experienced by the lower state is

$$U/\hbar \approx -\alpha\mathcal{E}_0^2/(4\hbar) = \frac{|\Omega_R|^2}{4\Delta_l}, \text{ (idealized two-level system)} \tag{5.12}$$

which exactly agrees with Eq. (3.46).

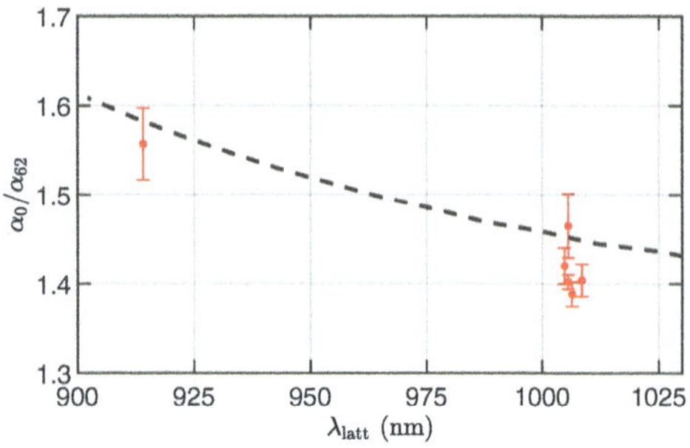

Fig. 5.1 The baseline ac polarizability ratio α_0/α_{62} at various lattice trap laser wavelengths. Red circles are experimental measurements. The black dotted line is the theoretical calculation. $X \to (1)1_u$ resonances are excluded from this plot for clarity of presentation but are included in the calculation

5.2 Lattice Clock Architecture

In one tremendously successful clock architecture, large numbers of neutral quantum absorbers are tightly confined in an optical lattice (see Fig. 1.1 for a caricature). Such lattice clocks, so far employing atomic optical transitions, have achieved unparalleled performance in both precision and accuracy. It is a near certainty that optical atomic lattice clocks will serve as primary frequency standards in the forthcoming redefinition of the SI second.

Clock accuracy and precision often advance in lockstep, closely intertwined in a virtuous circle. The foremost advantage of the lattice clock architecture over single- (or multi-) ion clock platforms is that the number of simultaneously interrogated quantum absorbers can be a few orders of magnitude larger, affording greatly reduced quantum projection noise [Eq. (1.1)]. Indeed, this was what motivated the development of lattice clocks two decades ago [5].

Optical trapping, however, is a double-edged sword. Generally, two distinct quantum states experience unequal light shifts in the presence of an external trapping laser due to their polarizability mismatch. For a lattice clock, this *differential* light shift presents at least two major complications. Firstly, the clock (transition) frequency would be light shifted from its unperturbed value, inhibiting accurate clock spectroscopy. Secondly, due to the thermal distribution of their speeds, the molecules sample spatially varying levels of laser intensity at different parts of the trapping potential, giving rise to the inhomogeneous broadening of the clock line (see Sect. 2.4.5), which diminishes the achievable precision.

The breakthrough came with the invention of the **magic wavelength** protocol. As the polarizability is predominantly a function of the applied laser frequency [Eq. 5.10], the key insight is that it is possible to find special crossings in the polarizabilities of the two clock states where the *differential* polarizability is essentially eliminated. By doing so, the internal degrees of freedom of the molecule become decoupled from their translational motion.

5.3 Molecular Magic Wavelength Protocol

5.3.1 Near-Resonant Magic Wavelengths

In general, the polarizabilities of two quantum states cross because the external laser is closer to the transition resonances of one of the states. This results in the disproportionate modification of its polarizability compared to that of the other state, thereby bridging the "baseline" polarizability difference [Eq. (5.14)]. Magic (or state-insensitive) trap design is non-trivial for molecules since the number of polarizability crossings, and the variety of resonances is much greater than in atoms.

For our prototypical molecular clock, we operate on the pure vibrational transition between the weakest bound and most tightly bound irrotational states in the $X^1\Sigma_g^+$ ground potential of $^{88}Sr_2$; $X(62, 0) \rightarrow X(0, 0)$. As a direct transition between $J = 0$ states is strictly forbidden at all orders of the multipole expansion, we drive the clock transition with a two-photon Raman process. The vibrational splitting of ~32 THz constitutes the clock frequency (f_{clock}). This choice of clock states is advantageous because f_{clock} is the largest possible pure vibrational frequency in the ground potential, and additionally, it provides direct access to the dissociation energy of the dimer (D_0), which is of prime interest to quantum chemists.

Magic trapping conditions can be engineered for $X^1\Sigma_g^+$ vibrational clock states by tuning the frequency of the lattice laser such that it off-resonantly addresses a transition from the more deeply bound clock state to a $J' = 1$ rovibronic state in $(1)1_u$, as illustrated in Fig. 4.7a. Figure 5.2a, b show the behavior of the differential light shift and the carrier linewidth as the lattice wavelength is swept across the $X \rightarrow 1_u$ transition over a relatively broad range.

As discussed in Sect. 2.2, being predominantly of triplet character, $(1)1_u$ states possess rather narrow natural linewidths; theoretically, $\Gamma_{1_u} \sim 2\pi \times 10\,\text{kHz}$. According to our spectroscopy (see Sect. 3.5, Table 3.6, and Fig. 3.14), deeply bound $(1)1_u$ states are well-isolated at intervals of ~2 THz. Under these conditions, a single $X \rightarrow 1_u$ resonance addressed by the lattice laser dominates the sum in Eq. (5.10). Consolidating the remaining contributions into a "background" term $\alpha_v^{\text{b.g.}} \equiv -\sum_{f\neq l} S_{fi}/(\hbar\Delta_f)$, we thus write

$$\alpha_v = -\frac{S_{li}}{\hbar\Delta_l} + \alpha_v^{\text{b.g.}}, \tag{5.13}$$

where $\Delta_f \equiv -(\omega_{fi} - \omega)$ is the angular frequency detuning of the lattice laser from the various rovibronic resonances, the subscript "l" is a shorthand label for the nearest $(1)1_u$ state off-resonantly addressed by the lattice, and we have neglected the counter-rotating terms $\propto 1/(\omega_{fi} + \omega)$ since the lattice laser angular frequency $\omega \sim 2\pi \times 300\,\text{THz}$; i.e., $1/(\omega_{fi} + \omega) \approx 0$.

For the Raman clock transitions investigated in this thesis, the polarizability of the deeply bound state α_v is strongly perturbed when the lattice is swept across an

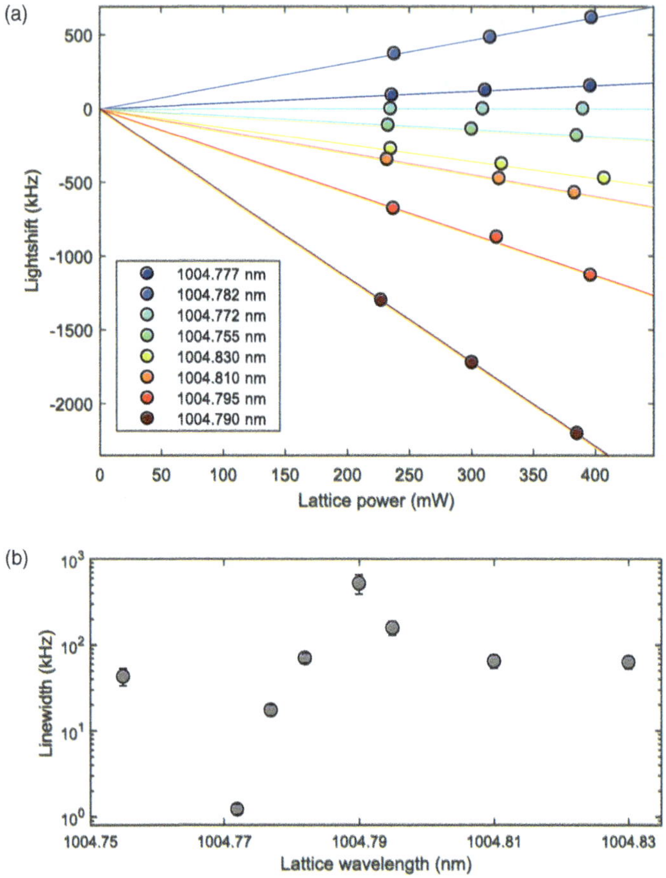

Fig. 5.2 Effect of lattice laser frequency on the $X(62, 0) \rightarrow X(0, 0)$ Raman clock. (**a**) Differential light shift of the clock transition versus lattice power, P_{latt}, at various lattice wavelengths (see legend). By off-resonantly addressing $X(0, 0) \rightarrow (1)1_u(9, 1)$, the differential light shift can be made to be largely insensitive to lattice power. For quick analysis, the line centers are obtained using Lorentzian fits to the spectroscopic resonances. Linear fits are made to the line centers as a function of P_{latt} with no fixed parameters, which accurately quantifies the slopes, L_0. Due to the asymmetrical carrier lineshape at non-magic wavelengths, fitting Lorentzian lineshapes systematically biases the vertical intercepts. In this plot, each set of raw line centers and their linear fit (color coded) are offset vertically so that the lines aesthetically intersect at the origin. (**b**) Dramatic narrowing of the clock linewidth by many orders of magnitude at the magic wavelength ($\approx$1004.7723 nm), measured at $P_{\text{latt}} \approx 300$ mW. Both (**a**) and (**b**) are prepared using the same raw dataset

$X \rightarrow 1_u$ resonance, while that of the weakly bound state α_{62} remains relatively unperturbed. By definition, the differential polarizability, $\alpha_v - \alpha_{62}$, is nulled at the **magic detuning** Δ_m. Therefore, we find that the "baseline" polarizability difference is

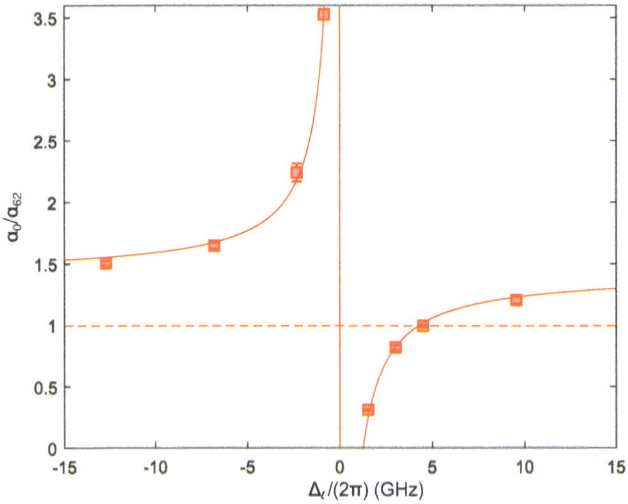

Fig. 5.3 Polarizability ratio α_0/α_{62} in the vicinity of $(1)1_u(9, 1)$. The "baseline" polarizability ratio, $\alpha_0^{\text{b.g.}}/\alpha_{62}$, is approximately 1.404(7) between 1004 and 1006 nm. The magic detuning occurs at 4.494(1) GHz from the $X(0, 0) \rightarrow (1)1_u(9, 1)$ resonance. Solid line is the dispersive fit in the form of $y = A/(x - x_0) + B$. Using Eq. (5.15), dispersive curves like this allow the $X \rightarrow 1_u$ transition strength to be accurately extracted using purely spectroscopic measurements. Due to good Franck-Condon overlap, $X(0, 0) \rightarrow (1)1_u(9, 1)$ is one of the strongest transitions of its kind, with $S_{li} = 1.335(35) \times 10^{-4}\,(ea_0)^2$. To obtain the transition strength, we make use of our polarizability model to estimate $\alpha_{62} \approx 484(10)$ a.u. between 1004 nm and 1006 nm. This figure is prepared using the same raw dataset as Fig. 5.2a, b

$$\alpha_v^{\text{b.g.}} - \alpha_{62} = \frac{1}{\hbar}\frac{S_{li}}{\Delta_m}. \tag{5.14}$$

Substituting this into Eq. (5.13), and computing the polarizability ratio α_v/α_{62}, we obtain

$$\frac{\alpha_v}{\alpha_{62}} = 1 + \frac{1}{\alpha_{62}}\frac{S_{li}}{\hbar}\left(\frac{1}{\Delta_m} - \frac{1}{\Delta_l}\right), \tag{5.15}$$

Figure 5.3 plots $\alpha_{v=0}/\alpha_{62}$ as a function of $\Delta_l/(2\pi)$, showing the characteristic $1/\Delta_l$ dispersive polarizability resonance feature. At the magic detuning ($\Delta_m = 2\pi \times 4.494(1)$ GHz for this pair of clock states and the chosen $(1)1_u$ rovibronic resonance), the polarizabilities of the clock states are matched and $\alpha_v/\alpha_{62} = 1$. The polarizability ratios are determined using Eq. (2.47) with the various L_0 obtained from the slope of the linear fits in Fig. 5.2a.

Taking the situation when $|\Delta_l| \gg |\Delta_m|$ in Eq. (5.15) such that $-S_{li}/(\hbar\Delta_l) \ll \alpha_v^{\text{b.g.}}$, we obtain the "baseline" polarizability ratio

$$\frac{\alpha_v^{\text{b.g.}}}{\alpha_{62}} \simeq 1 + \frac{1}{\alpha_{62}} \frac{S_{li}}{\hbar \Delta_m}, \tag{5.16}$$

and equating this with Eq. (2.47) under the same conditions

$$\frac{\alpha_v^{\text{b.g.}}}{\alpha_{62}} = 1 - \frac{4E_\text{r}}{h} \frac{L_0^{\text{b.g.}}}{f_\text{ax}^2/P_\text{latt}}, \tag{5.17}$$

we find a useful formula for the transition strength,

$$S_{li} \simeq \alpha_{62} \left(\frac{\alpha_v^{\text{b.g.}}}{\alpha_{62}} - 1 \right) \hbar \Delta_m = \alpha_{62} \left(\frac{-4E_\text{r} L_0^{\text{b.g.}}}{f_\text{ax}^2/P_\text{latt}} \right) \left(\frac{\Delta_m}{2\pi} \right). \tag{5.18}$$

Here, $L_0^{\text{b.g.}}$ is the "baseline" slope of the differential light shift (operationally, the lattice frequency is tuned to be somewhere in the middle of two adjacent $X \rightarrow 1_u$ resonances), and f_ax is the axial trapping frequency of α_{62} obtained from Raman sideband spectroscopy (see Sect. 2.4.4). This formula implies that for a given baseline polarizability mismatch between the clock states, the required magic detuning Δ_m *monotonically increases* with the $X \rightarrow 1_u$ transition strength. Importantly, the sensitivity of the clock transition to lattice frequency jitter is proportional to the *slope* of the lattice-induced light shift at the magic detuning, $-|\Omega_R|^2/(4\Delta_m^2)$ [Eq. (5.12)], and thus would *monotonically decrease* for larger transition strengths. In other words, near-resonant-type magic wavelengths based on stronger $X \rightarrow 1_u$ transitions place less stringent constraints on the required frequency stability of the lattice laser, and on the bandwidth of the spectral filter that suppresses amplified spontaneous emission (ASE) noise around the carrier. As we will see in the following subsection, $X(0, 0) \rightarrow (1)1_u(9, 1)$ is one of the strongest $X \rightarrow 1_u$ transitions below the $(1)0_u^+$ potential depth in $^{88}Sr_2$, making it ideally suited for engineering a near-resonant magic trap.

Finally, the close relationship between Eqs. (5.18), (5.2), (4.3) and (2.33) is concisely encapsulated by

$$|\Omega_R|^2 \simeq \left(\frac{\alpha_v^{\text{b.g.}}}{\alpha_{62}} - 1 \right) \frac{\omega_\text{ax}^2}{E_\text{r}/\hbar} \Delta_m, \tag{5.19}$$

which offers a convenient formula for converting between important magic trap-dependent, experimentally measured angular frequencies.

5.3.2 X to $(1)1_u$ Transition Strength Measurements

The previous subsection offers a great segue into a discussion of $X \rightarrow 1_u$ transition strengths. Recall that the polarizability of a dimer in the least bound state $X(62, 0)$ will be approximately twice that of an atom in $^1S_0$. Substituting $\alpha_{62} \simeq 2 \times \alpha_{\mathrm{Sr}}$ into Eq. (5.18), we get

$$S_{li} \simeq 2\alpha_{\mathrm{Sr}} \times \frac{2h^2}{M\lambda_{\mathrm{latt}}^2} \left(\frac{-L_0^{\mathrm{b.g.}}}{f_{\mathrm{ax}}^2/P_{\mathrm{latt}}} \right) \left(\frac{\Delta_m}{2\pi} \right), \tag{5.20}$$

where we remind the reader that M is the mass of the *dimer*, and λ_{latt} is the lattice wavelength at which $L_0^{\mathrm{b.g.}}$ and $f_{\mathrm{ax}}^2/P_{\mathrm{latt}}$ are measured.[6] The quantities appearing in Eq. (5.20) can either be measured very accurately (e.g., frequency measurements), or have existing values available from reliable literature (e.g., the polarizability of the Sr atom at various wavelengths, essentially enabling the calibration of the lattice intensity experienced by the molecules).

The transition strengths of various deeply bound $\mathrm{X}(v, 0) \rightarrow (1)1_u(v', 1)$ resonances obtained this way are listed in Table 5.1 and plotted in Fig. 5.4a, b. For $v = 4, 6 \rightarrow v' = 23$–$26$ in the 907–925 nm wavelength range, we use $\alpha_{\mathrm{Sr}} = 261.2$ a.u. at 914 nm from Ref. [20]. For the remaining $X \rightarrow 1_u$ transitions, we employ the molecular polarizability calculation delineated in Sect. 5.1.3 to estimate α_{62} at the various lattice wavelengths. Note that to convert polarizability in atomic units (a.u.) to MKS (SI) units, multiply the polarizability value in a.u. by $4\pi\epsilon_0 a_0^3$.

Transition strengths offer complementary insights into the chemical bonding of a molecule. As an illustrative example, we observe that the measured $X(6, 0) \rightarrow (1)1_u(v', 1)$ transition strengths exhibit decreasing trends in the two ranges explored by our experiment ($v' = 5$–7 and 23–26); see Fig. 5.4b. Calculations using the *ab initio* $(1)1_u$ potential [14], on the other hand, predict incompatible trends.

As alluded to in Sect. 3.5, it turns out that the *ab initio* $(1)1_u$ potential underestimated the depth of the potential by $\approx$300 cm^{-1}, merely 5% of the total depth [Fig. 3.16]. To arrive at this conclusion, our theory collaborators had the insight to model the short-range part of $(1)1_u$ with the simple Morse potential. The transition strengths with $X(6, 0)$ are calculated using the *ab initio* electronic transition dipole moment. The Morse $(1)1_u$ molecular wavefunctions are obtained numerically by solving the nuclear Schrödinger equation on an adaptive grid [21]. Similarly, the molecular wavefunction of $X(6, 0)$ is numerically calculated using the empirical ground potential from Ref. [16]. The physical origin for the observed trends in transition strength is the following. Since the classical turning points of the deeply bound X states are much further apart than those of $(1)1_u$, the Frank-Condon factors (see Sect. 4.2.1) largely depend on the spatial variation of the ground

[6] Ideally, this should also be within $\pm$1–2 nm of the $X \rightarrow 1_u$ transition wavelength, but care should be taken to avoid getting too close to a resonance or else Eq. (5.15) would not be valid.

Table 5.1 Transition strengths of deeply bound $(1)1_u(J' = 1)$ with deeply bound $X^1\Sigma_g^+(J = 0)$, given in units of $(ea_0)^2$. Values under the header "Skomorowski" are calculated using the unscaled *ab initio* potential [14], while those under "Majewska" are calculated using the Morse/Long-range potential [18]. The experiment values are determined using an accurate frequency-only spectroscopic method (see main text). For convenience, the *resonant* $X \rightarrow 1_u$ wavelengths are shown, rounded to 3 decimal places (see Table 3.6 for more precise values of $(1)1_u$ binding energies). We also list the requisite magic detunings for engineering a near-resonant-type magic trap via the respective $X \rightarrow 1_u$ transitions for various Raman clock pairs in $X^1\Sigma_g^+$. The initial clock state is $X(62, 0)$ in all but one entry (starred, $v = 6$, $v' = 25$ where it is $X(61, 0)$)

$X(v, 0)$	$(1)1_u(v', 1)$	$\lambda_{\text{latt}}^{\text{res.}}$ (nm)	$\frac{1}{2\pi}\Delta_m$ (GHz)	Skomorowski	Majewska	Exp. (this work)
8	13	1002.682	0.51(5)	4.8×10^{-5}	3.3×10^{-6}	$1.7(3) \times 10^{-5}$
6	5	1063.185	1.74(1)	1.0×10^{-4}	2.1×10^{-5}	$3.0(3) \times 10^{-5}$
6	6	1054.110	0.5(1)	7.8×10^{-6}	7.8×10^{-6}	$8(2) \times 10^{-6}$
6	7	1045.233	0.14(1)	2.9×10^{-5}	3.7×10^{-7}	$2.3(4) \times 10^{-6}$
6	23	925.973	1.380(2)	1.0×10^{-6}	2.1×10^{-5}	$4.3(2) \times 10^{-5}$
6	24	919.725	1.316(1)	2.8×10^{-5}	2.2×10^{-5}	$4.1(2) \times 10^{-5}$
6*	25*	913.597	0.67(2)	6.5×10^{-5}	1.3×10^{-5}	$2.1(1) \times 10^{-5}$
6	26	907.587	0.195(2)	7.5×10^{-5}	3.4×10^{-6}	$6.1(3) \times 10^{-6}$
4	11	1004.291	1.58(5)	2.9×10^{-5}	1.7×10^{-5}	$4.6(6) \times 10^{-5}$
4	25	907.636	2.30(4)	5.2×10^{-5}	3.9×10^{-5}	$6.7(3) \times 10^{-5}$
0	8	1012.903	4.22(5)	7.4×10^{-5}	5.7×10^{-5}	$1.2(2) \times 10^{-4}$
0	9	1004.787	4.494(1)	1.2×10^{-4}	6.2×10^{-5}	$1.34(4) \times 10^{-4}$
0	10	996.842	4.25(5)	1.6×10^{-4}	6.1×10^{-5}	$1.3(2) \times 10^{-4}$

state nuclear wavefunction. For $X(6, 0)$, the vibrational wavefunction resembles a Hermite-Gaussian polynomial with 6 nodes (c.f., quantum harmonic oscillator). Consequently, the transition strengths exhibit a characteristic zig-zag pattern when plotted against v'.

For completeness, we list the transition strengths for weakly bound $(1)1_u$ states with weakly bound $X^1\Sigma_g^+$ states in Table 5.2. These are measured using Autler-Townes spectroscopy, as described in Sect. 3.4.2. Additionally, Fig. 5.5 shows a heatmap of all $X(v, J = 0) \rightarrow (1)1_u(v', J' = 1)$ transition strengths calculated using the Morse/Long-range model (see Sect. 3.5.4).

5.3.3 *Clock State Lifetimes*

As a consequence of the Kramers–Kronig relations, a dispersive $(1/\Delta_l)$ frequency shift is necessarily accompanied by dissipating light scattering (or absorption). In our case, the absorption of lattice photons by the deeply bound clock state quenches its lifetime. According to Eq. (3.25), in the limit where the lattice laser detuning

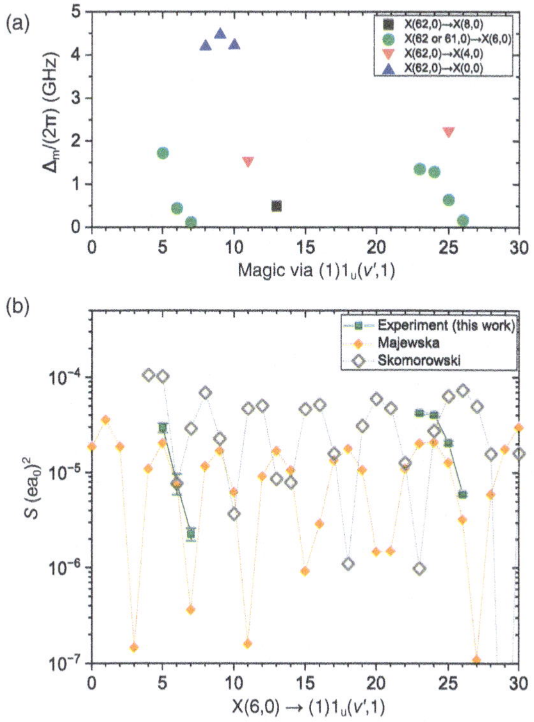

Fig. 5.4 Visualization of data from Table 5.1. (**a**) Magic detunings for various Raman transition pairs (see legend). Clock state-insensitive trapping conditions are created by tuning the lattice to off-resonantly address $(1)1_u(v', 1)$ from the deeply bound $X(v, 0)$ state. The magic wavelength at ≈1004.7723 nm, used for the vibrational clock transition $X(62, 0) \rightarrow X(0, 0)$ in this thesis, is 4.494(1) GHz blue detuned from $X(0, 0) \rightarrow (1)1_u(9, 1)$. It is the largest magic detuning of its kind hitherto observed in our experiment. (**b**) $X(6, 0) \rightarrow (1)1_u(v', 1)$ transition strengths versus v'. Green squares, measured; yellow diamonds, predicted using the Morse/Long-range potential [18]; gray open diamond, predicted using the *ab initio* potential [14]. Only the Morse/Long-range potential reproduces the observed relative trends in the transition strengths. As the *ab initio* $(1)1_u$ model underestimates the potential depth, its v'_{AI} labels are such that $v'_{\mathrm{AI}} = 0$ corresponds to $v' = 4$. Both theory calculations use the *ab initio* internuclear distance-dependent electronic transition moment; it is conceivable that scaling the electronic transition moment might bring the Morse/Long-range values to be in even closer agreement with the experiment

is much larger than the excited state linewidth ($\Delta_l \gg \Gamma_{1_u}$) and the intrinsic laser linewidth, the one-photon lattice scattering rate at the magic wavelength is

$$R_{1\gamma} \simeq \Gamma_{1_u} \frac{|\Omega_R|^2}{4\Delta_m^2}, \tag{5.21}$$

Table 5.2 Transition strengths of weakly bound $(1)1_u(J' = 1)$ with weakly bound $X^1\Sigma_g^+(J = 0)$, given in units of $(ea_0)^2$. Starred v' indicates a heavily Coriolis-mixed excited state. Headers have the same meaning as Table 5.1

$X(v, 0)$	$(1)1_u(v', 1)$	Skomorowski	Majewska	Exp. (this work)
62	−1	5.44×10^{-3}	4.56×10^{-3}	$5.53(8) \times 10^{-3}$
62	−2	3.6×10^{-4}	3.3×10^{-4}	$4.0(1) \times 10^{-4}$
61	−1	1.71×10^{-3}	1.68×10^{-3}	$1.74(3) \times 10^{-3}$
61	−2	6.95×10^{-3}	5.82×10^{-3}	$8.0(1) \times 10^{-3}$
60	−3*	1.32×10^{-2}	2.46×10^{-3}	$2.10(5) \times 10^{-3}$

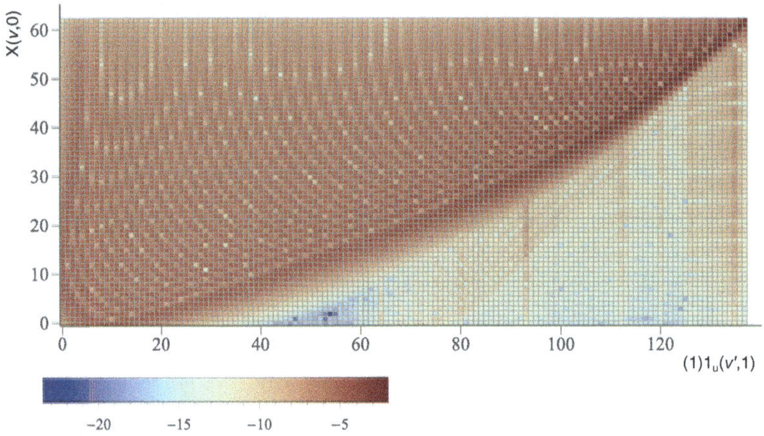

Fig. 5.5 Theoretical values of $X(v, 0) \to (1)1_u(v', 1)$ transition strengths computed using the Morse/Long-range potential from Ref. [18]. The scale on the colorbar is $\log_{10}[S/(ea_0)^2]$

which decreases quickly as $1/\Delta_m^2$ with greater magic detunings. Theoretically, if one-photon scattering is the sole[7] loss mechanism, then we should expect $X(0, 0)$ lifetimes that are slightly over 1 s at typical magic lattice trap depths.

It remains an outstanding conundrum that, empirically, the rate at which deeply bound molecules are lost in a near-resonant magic wavelength scales *quadratically* with lattice intensity ($\propto |\Omega_R|^4$), rather than linearly ($\propto |\Omega_R|^2$). The evidence for this effect on $X(0, 0)$ is shown in Fig. 5.6a, where Ω_R is the angular Rabi frequency of the $X(0, 0) \to (1)1_u(9, 1)$ transition off-resonantly addressed by the lattice. The etiology of the lattice-induced one-body losses is likely multifactorial.

One mechanism under investigation is that molecules in the deeply bound clock state are lost due to two-photon scattering (or photodissociation). Since deeply bound $(1)1_u$ states are of triplet character, it is plausible that a second lattice

[7] Note that there will always be two-body collisions that limit the lifetimes when more than 1 molecule is present at a lattice site. For a peak molecular density of 10^{11} cm^{-3} per lattice site and $k_2 \sim 10^{-10}$ cm^{3}s^{-1}, the molecular signal would quickly decrease within ∼ 100 ms. See Sect. 4.4.

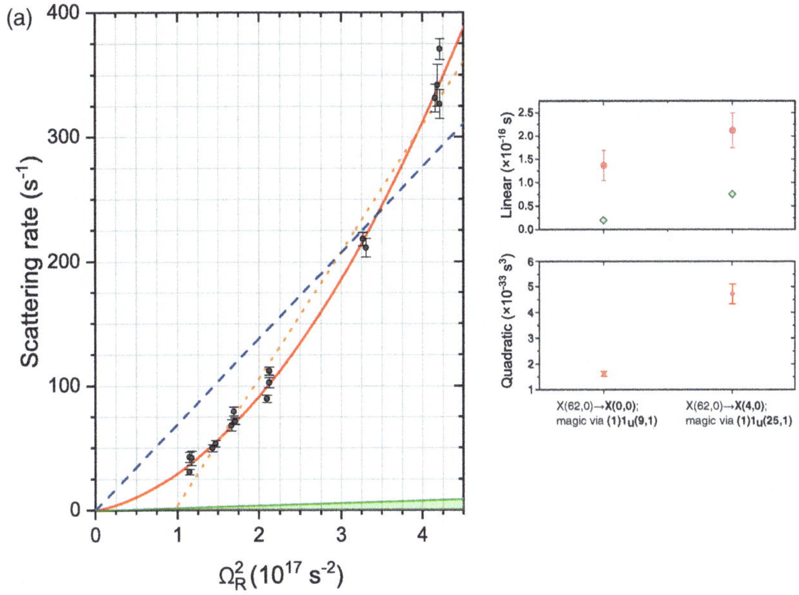

Fig. 5.6 (**a**) Lattice-induced scattering of $X(0,0)$ at the 1004.7723 nm magic wavelength for the vibrational Raman clock transition $X(62,0) \rightarrow X(0,0)$. Black circles; experimental data, consolidated over different methods of lattice frequency stabilization; e.g., phase-locked to frequency comb, piezo-locked to wavemeter, free-running. The scattering rates are computed as the inverse of the $X(0,0)$ state $1/e$ lifetimes (exponential loss, see Fig. 3.9 as an example). Here, a Ti:sapphire laser was used as the lattice laser source, with no additional optical filtering. The horizontal axis is the $X(0,0) \rightarrow (1)1_u(9,1)$ *angular* Rabi frequency, $|\Omega_R|^2$, calibrated from the dispersive light shift curves. Keeping the forward lattice beam power (P_{latt}) constant, this lattice-induced light shift in *angular frequency units* is $|\Omega_R|^2/(4\Delta_l)$; signs defined for the case where upleg is scanned [18, 22]. Δ_l is the *angular* frequency detuning of the lattice laser relative to the $X \rightarrow 1_u$ resonance. By fitting the dispersive feature to the form $y = A/(x - x_0) + B$ with at least 8 points (4 on each side of the $X \rightarrow 1_u$ resonance), we can extract the value of $|\Omega_R|^2 = 4A$ at the corresponding P_{latt}. Repeating this for at least 3 different P_{latt} yields a calibration curve for $|\Omega_R|^2$ versus P_{latt}. Red solid curve, parabolic fit with fixed intercept at origin; blue dashed line, linear fit with fixed intercept at origin; orange dotted line, linear fit with no fixed parameters; green shaded area, theoretical one-photon scattering rate $\Gamma_{1_u}|\Omega_R|^2/(4\Delta_m^2)$ for $\Gamma_{1_u} \leq 2\pi \times 10\,\text{kHz}$ and $\Delta_m = 2\pi \times 4.4937\,\text{GHz}$. The polynomial fits to the black data points are *not weighted* by the 1σ error bars, as they may be underestimated. (**b**) Parabola fit coefficients for $X(62,0) \rightarrow X(0,0)$ (magic via $(1)1_u(9,1)$) corresponding to the red curve in (**a**), and an analogous measurement with $X(62,0) \rightarrow X(4,0)$ (magic via $(1)1_u(25,1)$) in Ref. [18]. Values for the former are quoted in the main text. See also Table 5.1. Green diamonds in the top plot are the theoretical one-photon coefficients, $\Gamma_{1_u}/(4\Delta_m^2)$

photon mediates triplet-triplet transitions, resulting in two-photon dissociation. Molecular symmetry additionally constrains the contributing potentials to be gerade ($X0_g^+ \rightarrow (1)1_u \rightarrow g$). Starting from $X(0,0)$, two lattice photons would add $2 \times 1/(1004.7723\,\text{nm})$ wavenumbers and land 18843.43 cm^{-1} above the $^1S_0 + {}^1S_0$ threshold. For quick reference, the excitation wavenumbers of ground state atomic strontium are 14317.52 cm^{-1} ($^3P_0$), 14504.35 cm^{-1} ($^3P_1$), 14898.56 cm^{-1} ($^3P_2$), 18159.06 cm^{-1} ($^3D_1$), 18218.79 cm^{-1} ($^3D_2$), 18319.27 cm^{-1} ($^3D_3$), 20149.70 cm^{-1} ($^1D_2$), and 21698.48 cm^{-1} ($^1P_1$); values rounded from Ref. [23]. For the theoretical aspects and quantum chemistry calculations pertaining to this investigation, the reader is referred to Ref. [13]. Here, we approach the problem guided by our experimental measurements. We have not seen evidence of two-photon bound-to-bound transitions, and the losses are singly-peaked around the $X \rightarrow 1_u$ resonance [Fig. 5.7a]. The hypothetical second leg, therefore, would connect a bound-to-continuum transition from $(1)1_u$. Using Fermi's Golden rule, the rate of two-photon excitation is

$$R_{2\gamma} = \frac{\pi \hbar}{2} |\Omega_{2\gamma}|^2, \tag{5.22}$$

where $\Omega_{2\gamma}$ is the *effective* two-photon angular Rabi frequency involving the off-resonantly addressed $X \rightarrow 1_u$ transition as the first leg, and a hypothetical transition from the $(1)1_u$ state as the second leg (with angular Rabi frequency Ω_{hyp}, which itself is proportional to Ω_R because the same lattice laser excites both legs). Note that, here, $|\Omega_{2\gamma}|^2$ has dimensions of [Rate^2]/[Energy], or $1/(\text{J s}^2)$ in MKS (SI) units. From Eq. (3.36),

$$\Omega_{2\gamma} = \frac{\Omega_R \, \Omega_{\text{hyp}}}{2\Delta_m}. \tag{5.23}$$

Therefore, the lattice-induced two-photon excitation rate is

$$R_{2\gamma} = \frac{\pi \hbar}{8} \frac{|\Omega_R|^2 \, |\Omega_{\text{hyp}}|^2}{\Delta_m^2} \propto \frac{|\Omega_R|^4}{\Delta_m^2} \frac{S(1_\text{u} \rightarrow \text{hyp})}{S(\text{X} \rightarrow 1_\text{u})}, \tag{5.24}$$

where $S(1_\text{u} \rightarrow \text{hyp})$ is the transition strength of the hypothetical second leg, and $S(\text{X} \rightarrow 1_\text{u})$ is the $X \rightarrow 1_u$ transition strength (equivalent to S_{li} in the preceding subsections).

Figure 5.6b shows the linear and quadratic fit coefficients extracted by fitting the lattice-induced loss rate of the deeply bound states with a parabola (solid red curve in Fig. 5.6a). We compare these with an analogous measurement[8] involving the

[8] For $X(62, 0) \rightarrow X(4, 0)$ in Ref. [18], and $X(62, 0) \rightarrow X(6, 0)$ in Refs. [22, 24], the lattice light was derived from a tapered amplifier system. These previous studies utilize an off-the-shelf transmission grating to spatially separate ASE noise from the carrier over a long distance before the lattice light was coupled into a single-mode fiber to the experiment. Omission of this transmission grating severely quenches the deeply bound state lifetimes, even as far as $\sim$ 100 GHz detunings from the $X \rightarrow 1_u$ resonances [24]. This grating setup, however, has little to no effect on the ASE within $\lesssim$ 10 GHz. For data concerning the losses of the deeply bound state versus lattice intensity, only for $X(62, 0) \rightarrow X(4, 0)$ was the lattice additionally filtered through an optical cavity (to suppress near-carrier ASE), and were the lattice intensities calibrated in terms of $|\Omega_R|^2$. Hence, we omit the results from Ref. [22] in Fig. 5.6b.

$X(62, 0) \rightarrow X(4, 0)$ Raman transition, which has a magic wavelength at 907.6293 nm [18]. Heuristically, the baseline polarizability ratios $\alpha_0^{\text{b.g.}}/\alpha_{62}$ and $\alpha_4^{\text{b.g.}}/\alpha_{62}$ should be approximately equal since both deeply bound states lie near the bottom of the ground potential,[9] and the lattice wavelengths are not drastically dissimilar ($\approx$1005 nm versus $\approx$907 nm). Hence, the near-resonant light shift, $|\Omega_R|^2/(4\Delta_m)$, needed to bridge the baseline differential light shift is approximately equal for both cases.

The one-photon scattering rate [Eq. (5.21)] scales as $\sim |\Omega_R|^2/\Delta_m^2$. Since $\Gamma_{X\rightarrow 1_u}$ are similar, we expect that one-photon scattering will be commensurately smaller for larger magic detunings. The magic detunings under consideration differ by roughly a factor of 2 [Table 5.1]. Indeed, the linear loss coefficient for $X(0, 0)$ (made magic with $X(62, 0)$ via $(1)1_u(9, 1)$) is twice as small as that for $X(4, 0)$ (made magic with $X(62, 0)$ via $(1)1_u(25, 1)$). By comparison with Eq. (5.21), we extract $\Gamma_{1_u}^{\text{lin.}} = 2\pi \times 69(16)$ kHz, in tension with the theoretical expectation of $\sim 2\pi \times 10$ kHz.

We note that $X \rightarrow 1_u$ linewidths measured via direct one-photon spectroscopy [24] are also on the order of $\sim 2\pi \times 100$ kHz, but those experiments suffer from poor signal-to-noise as the sample of $X(6, 0)$ molecules were created using inefficient non-coherent transfer processes. It would be instructive in future work to revisit such direct spectroscopy using the large molecular samples of deeply bound states (and hence larger signal-to-noise) afforded by the STIRAP technique demonstrated in Chap. 4. Attention should be paid to possible broadening effects (e.g., thermal broadening due to non-magic lattice for $X \rightarrow 1_u$, magnetic fields, etc.).

The two-photon scattering rate to an excited, high-lying state [Eq. (5.24)] scales as $\sim |\Omega_R|^4/\Delta_m^2$. Therefore, any distinction (or lack thereof) in the two-photon scattering behavior would depend on how the ratios $\frac{S(1_u\rightarrow \text{hyp})}{S(X\rightarrow 1_u)}$ differ. Experimentally, we find that $X(0, 0)$ (made magic with $X(62, 0)$ via $(1)1_u(9, 1)$) has a quadratic coefficient that is approximately $3\times$ smaller than $X(4, 0)$ (made magic with $X(62, 0)$ via $(1)1_u(25, 1)$). Two-photon transitions are affected by the ellipticity of the lattice laser polarization (particularly if the states form a Ξ-system), possibly offering indirect verification of their role in the deeply bound state losses.

For a bound-to-continuum transition, we might imagine that $S(1_u \rightarrow \text{hyp})$ is fairly insensitive to the lattice wavelength. In this case, we should observe significantly faster scattering at near-resonant magic wavelengths engineered with smaller $S_{X\rightarrow 1_u}$. Confusingly, anecdotal evidence [22, 24] do not suggest a discernable dependence of the magic trap lifetimes of $X(6, 0)$ on $S_{X\rightarrow 1_u}$ in the 907–925 nm and 1045–1064 nm range, despite having varied $S_{X\rightarrow 1_u}$ by over an order of magnitude [Fig. 5.4b]. However, for these older experiments with $X(6, 0)$, the lattice intensities were not calibrated to $|\Omega_R|^2$, and the lattice light derives from a tapered amplifier system and the external optical filtering elements had to be laboriously realigned, posing the risk of uncontrolled ASE noise when the lattice wavelengths were changed. Moreover, the stated lattice wavelengths would result in two-photon

[9] Recall the famous result from elementary quantum mechanics that the polarizabilities of the vibrational states of a harmonic oscillator are independent of the vibrational quantum number.

transitions from $X(6,0)$ that closely address the $^1S_0 + {}^1D_2$ threshold where the assumption regarding the insensitivity of $S(1_\mathrm{u} \to \mathrm{hyp})$ to lattice wavelength may not be valid.

We next explore the frequency-dependence of the lattice-induced losses of $X(0,0)$ around the $X(0,0) \to (1)1_u(9,1)$ at a constant lattice intensity ($|\Omega_R|^2 \approx 1.8 \times 10^{17}\,\mathrm{s}^{-2}$, calibrated using dispersive clock light shift curves). Here, it is more appropriate to model one-photon scattering with the *saturated* excitation rate [25]

$$R_{\mathrm{sat}} = \frac{\Gamma_{1_u}}{4} \frac{|\Omega_R|^2}{|\Omega_R|^2/2 + \Delta_l^2 + \Gamma_{1_u}^2/4}. \tag{5.25}$$

We plot Eq. (5.25) versus Δ_l for two cases. In the first case, we use the value of $\Gamma_{1_u}^{\mathrm{lin.}}$ estimated from the linear coefficient of the parabolic fit above, shown as the dashed pink curve in Fig. 5.7a. In the second case, we use $\Gamma_{1_u} \lesssim 2\pi \times 10\,\mathrm{kHz}$ expected from theoretical calculations, depicted by the green shaded area in Fig. 5.7a. Neither of these curves comes close to the order of magnitude of our observed loss rates, once again ruling out one-photon scattering as the sole mechanism. Without strong physical motivation, we fit Eq. (5.25) plus a constant background term to the loss rates (solid red curve in Fig. 5.7a), extracting $\Gamma_{1_u} = 2\pi \times 120(10)\,\mathrm{kHz}$.

The measurements in Fig. 5.7a exhibit asymmetric features that warrant discussion. The losses are consistently higher on the blue side of the $X \to 1_u$ resonance, around $\Delta_l \approx 2\pi \times 1$–$2\,\mathrm{GHz}$. This could possibly be an artifact of our measurement scheme. To obtain the loss rates, we perform STIRAP (see Chap. 4) to produce a sample of $X(0,0)$ molecules before finally reversing the STIRAP process to recover $X(62,0)$ that we can detect. By varying the duration between the forward and reverse STIRAP pulses [Fig. 5.7c], we hold the $X(0,0)$ in the lattice for a variable time, t. We monitor the exponential decay of the molecule number after a roundtrip STIRAP versus t, and estimate the lattice-induced loss rates as the inverse of the $1/e$ time constant. It turns out that the **tune-out detuning** (Δ_{to}) for $X(0,0)$ occurs at approximately $2\pi \times 1.29(1)\,\mathrm{GHz}$. At the tune-out wavelength, $\alpha_0 = 0$, and $X(0,0)$ molecules do not experience any optical dipole force from the trap. In fact, it can be shown using Eqs. (5.13) and (5.15) that

$$\frac{\alpha_v^{\mathrm{b.g.}}}{\alpha_{62}} = \frac{1}{1 - \Delta_{\mathrm{to}}/\Delta_m}. \tag{5.26}$$

Therefore, for $0 \leq \Delta_l < \Delta_{\mathrm{to}}$, $X(0,0)$ molecules are *repelled* out of the trap, which would artificially exaggerate the losses. Due to practical limitations, the roundtrip STIRAP transfer times are at least 100–200 μs. To investigate the exceedingly fast loss rates within a very narrow band ($\sim 1\,\mathrm{GHz}$) of the $X \to 1_u$ transition, we adopt a different timing strategy shown in Fig. 5.7d. Here, we fix the roundtrip STIRAP duration, perform STIRAP in free flight, and vary the time the $X(0,0)$ molecules are exposed to lattice light during the roundtrip. Fast pulsing of the lattice

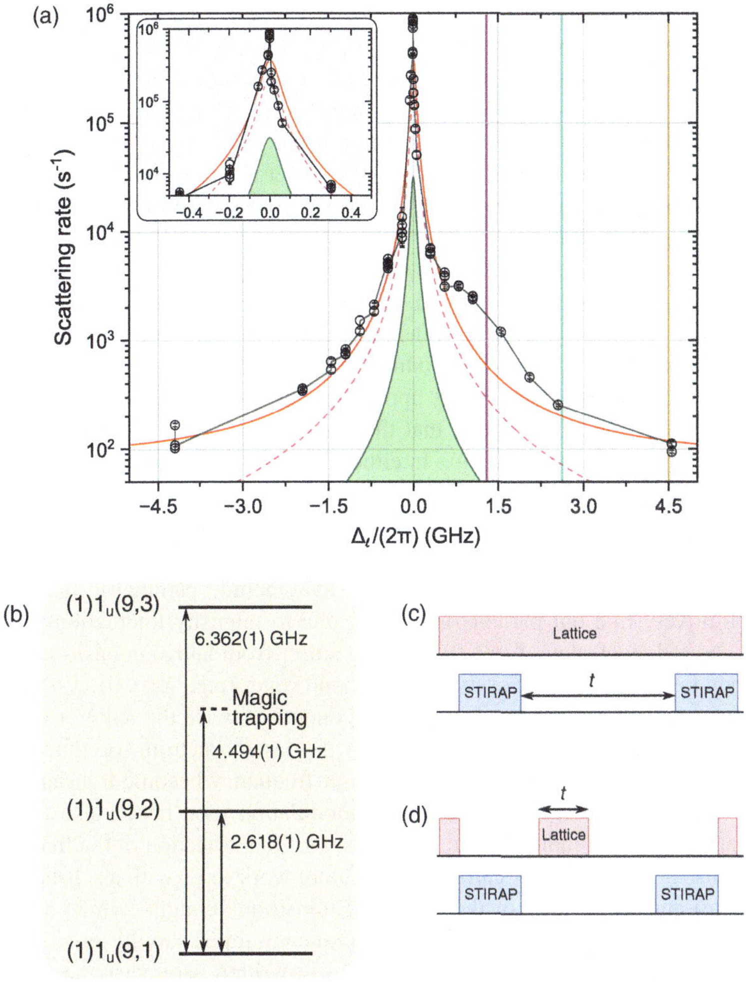

Fig. 5.7 (**a**) Frequency dependence of the lattice-induced losses; $|\Omega_R|^2 \approx 1.8 \times 10^{17}\,\mathrm{s}^{-2}$. Black points, the inverse of the $1/e$ exponential decay time constants. Connecting lines are guides to the eye. Inset shows a zoomed view spanning 1 GHz around the $X(0, 0) \to (1)1_u(9, 1)$ resonance. See the main text for a discussion of the asymmetrical features and the plotted peak functions. Vertical colored lines mark physically significant detunings: Magenta, the tune-out detuning where $\alpha_0 = 0$; cyan, $X \to 1_u(J' = 2)$; gold, the magic detuning where $\alpha_0 = \alpha_{62}$. (**b**) Rotational splittings of $(1)1_u(9, J')$; see also Table 3.6. (**c, d**) Experimental timing sequences for the measurements in (**a**)

light is accomplished using an acousto-optic modulator.[10] As shown by the inset to Fig. 5.7a, the losses appear to have the opposite behavior, where blue detunings result in smaller loss rates. One hypothesis is that although blue detunings repel the molecules away from high-intensity regions of the pulsed lattice beam, this only occurs for a very short duration in free flight, and most of the molecules are recaptured by the lattice. Hence, we would detect a larger number of remaining $X(0, 0)$ molecules because the average intensity experienced by the molecules over the same time is smaller.

The magic wavelengths are rather close to the positions of $X(v, J = 0) \rightarrow 1_u(v', J' = 2)$ transitions [Fig. 5.7b]. Exotic mechanisms might open up this loss channel; e.g., electric quadrupole transitions due to the tight focusing of the lattice laser beam, or lattice-induced admixture of $J' = 1$ and $J' = 2$ states of $(1)1_u$ (previously observed under strong applied magnetic fields [26]). For the latter, we may expect, to leading order, that the degree of $J' = 1, 2$ state admixture is proportional to the lattice intensity. In either case, the overall scattering rate due to $X(J = 0) \rightarrow 1_u(J' = 2)$ would scale quadratically with lattice intensity. The lack of a sharp resonance in Fig. 5.7a at the expected location for a $X \rightarrow 1_u(J' = 2)$ transition might rule out this possibility.

Known technical effects that cause trap loss include parametric heating [27]. We definitively rule out parametric heating due to intensity fluctuations since the weakly bound clock state $X(62, 0)$ does not suffer from lattice intensity-dependent one-body losses [Fig. 3.9]. For the deeply bound states (e.g., $X(0, 0)$, $X(4, 0)$, etc.), there is a realistic chance that Fourier components at twice the axial or radial trap frequencies (e.g., occurring intrinsically in the laser spectrum, or introduced by the phase-locked loop, etc.) in the lattice laser frequency become translated by the $X \rightarrow 1_u$ resonance into a polarizability modulation (and hence also trap depth modulation), thereby inducing parametric losses. Quantification of the frequency or phase noise around the laser carrier line in future work (e.g., with a self-heterodyne technique) and knowledge of the $X \rightarrow 1_u$ transition strength (which determines the sensitivity of the magic trap depth to frequency jitter) should provide a rough estimate of the extent to which frequency-noise-written parametric heating might account for the observed clock state lifetimes.

To conclude our discussion of clock state lifetimes, we show the limit that blackbody radiation (BBR) poses to the lifetimes under a room-temperature setting in Fig. 5.8. Our theory collaborators made these calculations, and the reader is directed to Ref. [13] for a rigorous account. The calculations favorably suggest lifetimes upward of 10^5 years. Indeed, infrared inactivity is one of the advantages homonuclear molecules offer for precision measurement applications. While current technical limitations (such as lattice scattering or two-body losses) render these

[10] Care should be taken to eliminate (if not minimize) leakage light from other diffraction orders. In practice, due to internal crystal reflections, etc., some leakage inevitably gets sent to the experiment. As the lattice intensity is very high, spurious spectroscopic features can appear relative to the main carrier resonance.

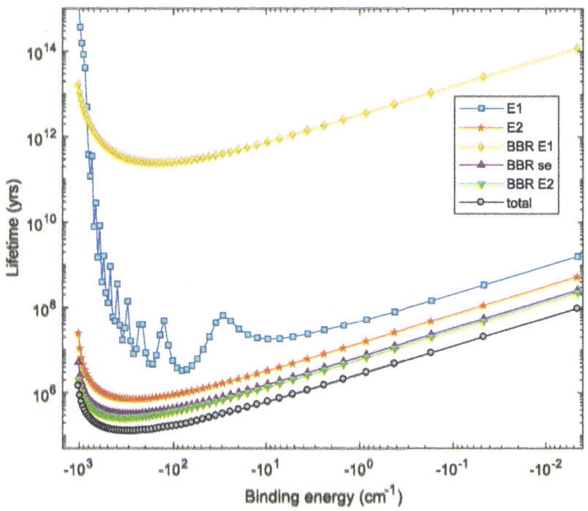

Fig. 5.8 Theoretically calculated BBR-limited lifetimes of $X^1\Sigma_g^+(v, J = 0)$. At a surrounding temperature of 300 K, the lifetimes are largely quenched by electric quadruple ($E2$) transitions to $J = 2$ states mediated by BBR photons. Reproduced with permission from Refs. [13, 22]; full citation: Kondov, S.S., Lee, C-H., Leung, K.H. et al. Molecular lattice clock with long vibrational coherence. *Nat. Phys.* **15**, 1118–1122 (2019)

tremendously long BBR lifetimes inconsequential, one may envision a future where molecular quantum science progresses to a stage that would allow harnessing the rich rovibrational structure of molecular states for metrology and quantum information applications.

5.4 Vibrational Molecular Clock

5.4.1 Clock Scheme and Fiber Noise Cancellation

The two clock lasers are detuned from the intermediate excited state $(1)0_u^+(11, 1)$. This pathway through a deeply bound $(1)0_u^+$ state offers favorable Rabi frequencies, and we have seen how it facilitated stimulated Raman adiabatic passage (STIRAP) transfer between $X(62, 0)$ to $X(0, 0)$ described in Chap. 4. By contrast, weakly bound $(1)0_u^+$ states near the intercombination line (which we utilized in Ref. [22]) are expected to have negligibly small transition strengths to $X(0, 0)$ due to poor Franck-Condon overlap [18, 28]. The relevant potentials are shown in Fig. 5.9a.

The clock measurements take place in a retroreflected 1D optical lattice at $\approx$1004.7723 nm, oriented horizontally with respect to gravity due to geometric limitations in this study. The lattice overlaps with the atom cloud throughout the atomic cooling sequence ($\sim$500 ms), and atoms with kinetic energies lower than the trap depth are loaded into the lattice. The surrounding magnetic field is lowered to $<$0.6 G to prepare for molecule production and spectroscopy.

Trapped samples of ultracold molecules are created by photoassociating laser cooled strontium atoms at 2 μK to $(1)1_u(-1, 1)$. This efficiently produces $X(62, 0)$ ground state molecules thanks to the large transition strength [18]. To purify the rotational population of the molecular gas, we photodissociate the $X(62, 2)$ molecules 30 MHz above the $^1S_0 + {}^3P_1$ threshold, imparting more than sufficient kinetic

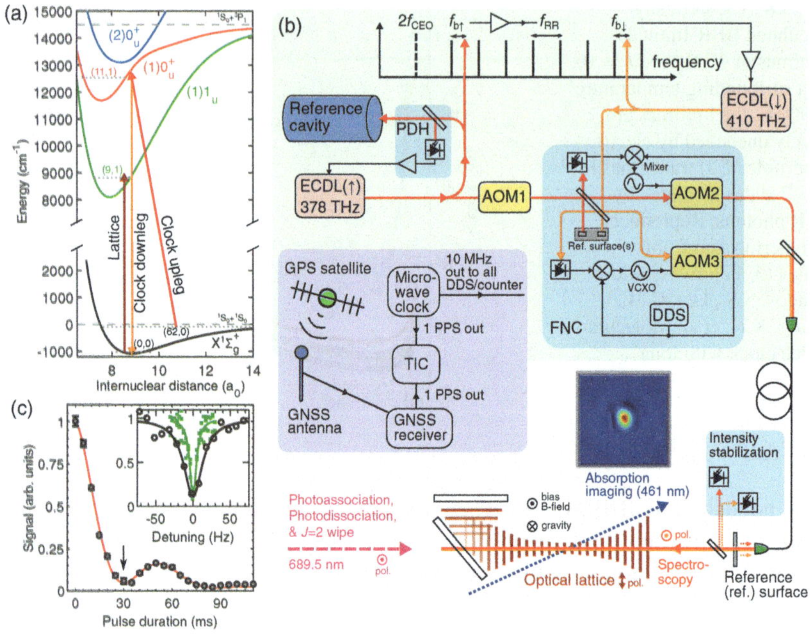

Fig. 5.9 Vibrational molecular lattice clock. (**a**) Raman lasers (upleg, red arrow; downleg, orange arrow) detuned from an intermediate state in $(1)0_u^+$ probe the two-photon vibrational clock transition between ($v = 62$, $J = 0$) and ($v = 0$, $J = 0$) in the $X^1\Sigma_g^+$ ground potential. The optical lattice (brown arrow) off-resonantly addresses an isolated rovibronic state in $(1)1_u$ to induce magic trapping conditions. (**b**) Experimental setup. The upleg master laser is stabilized to a reference cavity using the Pound-Drever-Hall (PDH) technique, and its phase coherence is transferred to the downleg laser via a frequency comb. The molecules are held in the 1D optical lattice. Co-propagating clock lasers are delivered to the molecules via an optical fiber with active fiber noise cancellation (FNC). The spectroscopic signal derives from absorption imaging of $X(62, 0)$ photofragments at a slight grazing angle relative to the lattice. A rubidium microwave standard acts as a flywheel oscillator, linking the molecular clock to GPS time for the absolute frequency measurement. (**c**) Two-photon Rabi oscillations between the clock states driven at the operational probe intensities (filled circles, experimental data averaged over 8 consecutive runs, error bars represent 1σ uncertainties; solid red line, analytical fit to an exponentially decaying sinusoid). We observe lines as narrow as 11(1) Hz (inset, green squares). For clock operation, we perform Rabi spectroscopy with a 30 ms π-pulse duration (indicated by the black arrow), resolving 30(2) Hz linewidths consistent with the expected Fourier limit (inset, black open circles). Each point in the inset is a single shot of the experiment, and solid lines are Lorentzian fits

energy to guarantee these photofragments leave the trap. We do this concurrently with PA (pulse durations of ~2 ms). Here, the binding energy of $(1)1_u(1, -1)$ with respect to $^1S_0 + {}^3P_1$ and that of $X(62, 2)$ with respect to $^1S_0 + {}^1S_0$ fortuitously add to be ≈420 MHz, which is the frequency shift introduced by double-passing the first-order diffraction of an AOM with center frequency of 210 MHz, which we often have in surplus in the lab. Therefore, rather than utilizing a separate phase-locked laser to wipe away $J = 2$ molecules, we simply add a frequency sideband to the PA laser with an AOM. We lock the PA laser so that the unperturbed frequency ({0, 0} order) addresses the PA transition $^1S_0 + {}^1S_0 \rightarrow (1)1_u(-1, 1)$, and the {+1, +1} order (retroreflected by a D-shaped mirror) addresses $X(62, 2) \rightarrow {}^1S_0 + {}^3P_1$. Both orders pass through the same quarter waveplate for retroreflection, are recombined at a polarizing beamsplitter placed before this AOM. These PA and $J = 2$ wipe beams are then combined with the $X(62, 0)$ state detection (i.e. photodissociation) beam. Finally, all three ~689 nm laser beams are injected into a single polarization-maintaining fiber and sent to the experiment table. The remaining atoms after PA are wiped out of the trap with resonant 461 nm laser light. As the state-selective photofragmentation of $X(62, 0)$ and subsequent absorption imaging of the slow-moving atoms destroys the prepared molecular sample, the entire sequence has to be iterated to scan the clock transition.[11] For more details on state preparation and detection, see Chap. 2.

Raman clock spectroscopy is deeply in the Lamb-Dicke regime for co-propagating probes along the axial direction of the optical lattice (Lamb-Dicke parameter $\eta_{\mathrm{ax}} \lesssim 0.02$, see also Sect. 2.4.4). As discussed in Sect. 3.1.2, the upleg master clock laser at 378 THz (793 nm) is stabilized to a high finesse ultra-low expansion reference cavity. The phase coherence of the upleg is transferred to the teeth of an erbium-fiber-laser-based optical frequency comb by actuating on its repetition frequency [Fig. 5.9b]. The downleg clock laser at 410 THz (731 nm) is phase locked to the comb, thereby inheriting the phase stability of the upleg. The carrier-envelope offset frequency of the comb is stabilized to a rubidium standard that also serves as the laboratory timebase. More details on the high finesse cavity, frequency comb, and rubidium timebase can be found in Chap. 3.

Since the Raman transition samples the correlated frequency difference of the clock lasers, frequency drift and spectral broadening due to the instability and linewidth, respectively, of the upleg master laser are greatly suppressed. As mentioned in Chap. 3, observations of the counted comb repetition rate against the rubidium timebase actively steered by a GPS disciplined oscillator for several months prior to the clock campaign reveal a cavity drift rate of 30 mHz/s, which we compensate using a linearly-ramped AOM in the optical path of the master laser to the cavity. The frequency synthesizer that performs this linear feedforward compensation updates every second. The residual linear drift of the master laser due to imperfect feedforward is approximately 3 mHz/s during the campaign, consistent with the observed drift of the molecular clock line centers over the same period, after accounting for the comb teeth difference [$3\,\mathrm{mHz/s} \times (1 - N_\uparrow/N_\downarrow) \approx 0.2\,\mathrm{mHz/s}$].

The upleg is passed through an acousto-optic modulator (AOM1 in Fig. 5.9b), and the first order diffraction ({+1}) is used to iteratively step the difference frequency of the clock lasers across f_{clock}. AOM1 controls the interrogation duration by pulsing the upleg, and we leave the downleg constantly irradiating (but blocked with a mechanical shutter during the state preparation process).

[11] The stability of the clock is therefore susceptible to degradation via the Dick effect.

The clock lasers are injected into the same polarization-maintaining single-mode fiber (enclosed in a 1" thick flexible rubber foam pipe along its entire length for passive insulation from the environment) and delivered to the adjacent optical table where the experiments take place. To minimize the number of optical elements and unstabilized path lengths, the clock lasers interrogate the molecules from the opposite direction as the PA and photodissociation (~689 nm) lasers. Since the clock laser wavelengths are sufficiently different that the laser beams may sample non-identical paths in a given refractive medium, active **fiber noise cancellation** (FNC) [29, 30] on each clock leg is implemented using independent phase actuators (acousto-optic modulators AOM2 and AOM3 in Fig. 5.9b). An exploded view of the FNC scheme for one of the clock lasers is shown in Fig. 5.10; the scheme is duplicated for the other clock laser. If laser power is critical for future applications, the 50:50 non-polarizing beam splitter (BS) may be substituted with a polarizing beam splitter and a Faraday rotator [31]. The voltage-controlled crystal oscillators[12] (VCXO) provide the RF frequencies for AOM2 and AOM3. The optical phase reference surface at the experiment table is a single partially reflecting mirror, while the surfaces on the laser table are mounted on a common rigid pedestal post with the clock lasers approaching the surfaces in the same direction. The total uncompensated path in air is approximately 50 cm.

An achromatic waveplate maximizes the power through a Glan-Thompson prism (which serves to clean up the polarizations of the clock laser beams) placed after the phase reference surface. Immediately after the prism, the clock laser beams are separated by a high-quality (long-pass) dichroic beam splitter, and a small portion of their light is picked off. To avoid contamination from the other leg due to leakage from the dichroic, the picked off beams are further reflected off dichroics (instead of regular mirrors) before being measured on separate photodiodes.[13] Independent intensity stabilization of each clock leg is done by actuating on the RF powers to AOM2 and AOM3. The integrator on the servo for the upleg power is disabled (with a TTL signal from the experiment control sequence) during the short duration for state preparation (i.e., PA, $J = 2$ and atom wipe), when its light is pulsed off by AOM1. This keeps the servo output within range (i.e., not railed) so that the power is able to quickly return to the setpoint when the light is pulsed on again for clock interrogation. The separated clock laser beams can be individually attenuated by separate neutral density filters mounted on motorized holders. This is preferred to changing the setpoint of the intensity servos as the latter would affect the robustness of the FNC if insufficient clock laser light were coupled to the experiment table. Finally, the clock laser beams are recombined on a high-quality dichroic, and

[12] Crystek 744-CRBSVS-02-80.000-ND.

[13] The alternative is to put a narrow bandpass filter before each photodiode. However, we found that this introduces a stubborn etalon-like effect that causes the measured laser power to vary with slight changes in the laser alignment to the photodiode, even when the surface of the filter is angled with respect to the photodiode sensor.

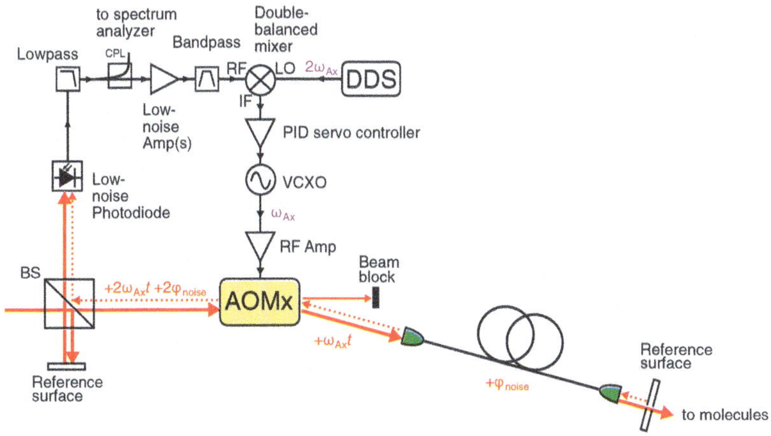

Fig. 5.10 Fiber noise cancellation (FNC) with an optoelectronic phase-locked loop. A laser beam is initially split by a 50:50 non-polarizing beam splitter (BS), and one arm is retroreflected by a phase reference surface on the experiment table toward a low-noise photodiode. The other arm is diffracted by an acousto-optic modulator (AOMx) driven by a voltage-controlled crystal oscillator (VCXO), which has exceptional phase purity. The laser is shifted by angular frequency ω_{Ax} so that the phase of the laser field evolves with an *additional* phase $\omega_{\mathrm{Ax}}t$ over time t. Then, the diffracted order is injected into an optical fiber to the experiment table. Passage through the fiber writes phase noise ϕ_{noise}. To detect this, the laser is retroreflected back through the fiber by a phase reference surface at the experiment table and double-passed through AOMx. Heterodyne beat detection at the low-noise photodiode thus obtains the phase noise $2\phi_{\mathrm{noise}}$ centered at angular frequency $2\omega_{\mathrm{Ax}}$. This beat note is mixed down to dc with a stable RF reference from a direct digital synthesizer, and the intermediate frequency (IF) output of the mixer is *proportional* to the phase error ϕ_{noise}. A servo controller shapes this error signal and actuates on the VCXO to write an additional phase $-\phi_{noise}$ via AOMx to preemptively cancel the fiber phase noise. The entire detection process occurs at the speed of light, which is faster than the rate at which ϕ_{noise} changes due to perturbations in a typical lab environment. Ultimately, the optical phases at the two reference surfaces are made equal by the phase-locked loop; i.e., laser light *after* the reference surface at the experiment table is fiber noise canceled. Without FNC, linewidth broadening due to fiber phase noise is typically $\lesssim$ 100 Hz on an optical laser field

further combined with the lattice on another dichroic.[14] The clock laser beams are coaligned to the forward pass lattice beam over several meters across the room, and are visually verified to overlap at the focus of the latter on an inexpensive CMOS or CCD camera. The polarizations of the probes are identical, linear, and parallel to the small applied magnetic field, but perpendicular relative to that of the lattice in this work. As a reminder, for a Raman transition involving $J = 0 \leftrightarrow J' = 1$, the two-photon Rabi frequency is maximized when the upleg and downleg have identical polarizations, and vanishes if orthogonal (see Sect. 3.4).

Figure 5.9c shows two-photon Rabi oscillations driven by the clock lasers at the operational Rabi frequencies (see also Sect. 3.4 and Fig. 3.9). We may lower the

[14] 805 nm short-pass.

clock laser beam powers further and use pulse durations of $\sim$100 ms, producing clock lines with full width at half-maximum (FWHM) as narrow as 11(1) Hz corresponding to a Q-factor of 2.9×10^{12} (solid green squares in the inset to Fig. 5.9c). As mentioned in Sect. 5.3.3, in the absence of other fields, the BBR limited lifetimes of the clock states exceed 10^5 years in a room temperature environment [13, 22], suggesting no fundamental limit for the Q-factor. Nevertheless, the main technical challenges in the current iteration of the molecular clock are two-body molecular losses close to the universal rate (see Sect. 4.4) and lattice light-induced one-body losses for $X(0, 0)$ that scale quadratically with the trap depth (see Sect. 5.3.3). At the operational density and lattice trap depth in this work, these losses quench the spectroscopic signal fast enough that the molecular densities vary significantly over pulse durations of $\gtrsim$60 ms. Therefore, as a compromise, we evaluate clock systematics by performing Rabi spectroscopy with a 30 ms π-pulse, scanning Fourier-limited peaks of 30(2) Hz (black open circles in the inset to Fig. 5.9c). A typical spectrum consists of 15 experimental iterations (taking a total duration of $\sim$20 s), from which we determine the line center by fitting a Lorentzian function in post process.

5.4.2 *Basic Concepts from Statistics, and the Frequency Measurement Chain*

Often, we have to extrapolate the measured clock line centers to determine the unperturbed clock frequency. This involves curve fitting—the mathematics of which is well established and exhaustive accounts can be found in Refs. [32–35]. Below, we provide a brief overview to establish the notation and language. Suppose we have n number of observations of the dependent variable $\mathbf{y} = \{y_1, y_2, \ldots, y_n\}$ with estimated error bars $\boldsymbol{\sigma} = \{\sigma_1, \sigma_2, \ldots\}$, measured at the values of the independent variable $\mathbf{x} = \{x_1, x_2, \ldots, x_n\}$. While there is minimal risk of confusion given that the context is clear, we explicitly point out that the symbols "y" and "x" in this subsection (and the following ones) do not represent the spatial coordinate axes labels. For example, $\mathbf{y}$ could be the clock shift Δf_{clock} or the fitted values of f_{A1} that correspond to the clock line centers, and $\mathbf{x}$ could be an experimental parameter such as trap depth or the iteration number of a consecutive series of measurements. We model the behavior of the dependent variable versus the independent variable with the function $f_{\text{model}}(\mathbf{x}, \boldsymbol{\Theta})$, which has p number of fit parameters $\boldsymbol{\Theta} = \{\Theta_1, \ldots, \Theta_p\}$. Taking $1/\sigma_i^2$ as the weights for fitting, the least-squares algorithm finds the set of fit parameters, $\boldsymbol{\Theta}_0$, that minimizes the chi-square

$$\chi^2 \equiv \sum_{i=1}^{n} \left(\frac{y_i - f_{\text{model}}(x_i, \boldsymbol{\Theta})}{\sigma_i} \right)^2 . \tag{5.27}$$

To estimate the errors for the fit parameters, it is customary to compute the $n \times p$ matrix $\mathbf{A}$ where

$$A_{ij} \equiv \frac{1}{\sigma_i} \frac{\partial f_{\mathrm{model}}(x_i, \mathbf{\Theta})}{\partial \Theta_j}\bigg|_{\mathbf{\Theta}_0}. \tag{5.28}$$

The standard error of the fit parameter Θ_i is then

$$\sigma_{\Theta_i} \approx \sqrt{\left[(\mathbf{A}^{\mathsf{T}}\,\mathbf{A})^{-1}\right]_{ii}}, \tag{5.29}$$

where $\mathbf{A}^{\mathsf{T}}$ is the transpose of matrix $\mathbf{A}$, and $()^{-1}$ indicates a matrix inversion. The *reduced* chi-square is the chi-square divided by the number of degrees of freedom in the fit,

$$\chi^2_{\mathrm{red}} = \frac{\chi^2}{n-p}. \tag{5.30}$$

Occasionally, the data error bars σ_i may be underestimated, and the dataset is overscattered (i.e., $\chi^2_{\mathrm{red}} > 1$). In such situations, it is considered more appropriate to scale up the standard error by $\sqrt{\chi^2_{\mathrm{red}}}$, and quote the revised (or scale corrected) standard error

$$\tilde{\sigma}_{\Theta_i} = \sigma_{\Theta_i} \times \sqrt{\chi^2_{\mathrm{red}}}. \tag{5.31}$$

Note that $\mathbf{C} \equiv (\mathbf{A}^{\mathsf{T}}\,\mathbf{A})^{-1}\chi^2_{\mathrm{red}}$ is called the *covariance matrix*. Its inverse, $\mathbf{H} = \mathbf{C}^{-1}$, is also known as the *Hessian*. In many curve-fitting software packages, the revised standard errors are quoted by default, and they are calculated as the square root of the diagonal elements of $\mathbf{C}$ after multiple numerical iterations to minimize χ^2; i.e., $\tilde{\sigma}_{\Theta_i} \simeq \sqrt{C_{ii}}$.

As an illustrative example, let us consider the weighted average. Here, the fit function consists of just a single parameter; i.e, $p = 1$, and $f_{\mathrm{model}}(\mathbf{x}, \mathbf{\Theta}) = \Theta$. We minimize χ^2 by setting $\partial\chi^2/\partial\Theta = 0$ and find $\Theta_0 = (\sum_i y_i/\sigma_i^2)/(\sum_i 1/\sigma_i^2)$, which is the formula for the weighted average. It is similarly straightforward to show that $(\mathbf{A}^{\mathsf{T}}\,\mathbf{A})^{-1} = 1/\left(\sum_i 1/\sigma_i^2\right)$. Therefore, the revised standard error of the weighted mean is $\tilde{\sigma}_{\Theta} = \sqrt{1/\left(\sum_i 1/\sigma_i^2\right)} \times \sqrt{\chi^2_{\mathrm{red}}}$, if $\chi^2_{\mathrm{red}} > 1$.

The frequency chain [Fig. 5.9b] implies that the upleg laser frequency at the molecules is

$$f_{\uparrow}^{\mathrm{mol}} = N_{\uparrow} f_{\mathrm{RR}} + 2 f_{\mathrm{CEO}} + f_{\mathrm{b}\uparrow} + f_{\mathrm{A1}} + f_{\mathrm{A2}}, \tag{5.32}$$

where $2 f_{\mathrm{CEO}}$ is the frequency of the zeroth comb tooth after sum frequency generation (i.e., f_{CEO} is the carrier-envelope offset of the *fundamental* output of the comb at 1560 nm, see Sect. 3.1.2), $f_{\mathrm{b}\uparrow}$ is the beat frequency between the laser at the comb and the $N_\uparrow$-th comb tooth that we phase lock to. f_{A1} and f_{A2} are the frequency shifts introduced by AOM1 and AOM2 respectively. Similarly, the downleg laser frequency at the molecules is

$$f_\downarrow^{\mathrm{mol}} = N_\downarrow f_{\mathrm{RR}} + 2f_{\mathrm{CEO}} + f_{\mathrm{b}\downarrow} + f_{\mathrm{A3}}. \tag{5.33}$$

To suppress relative phase fluctuations due to the comb carrier-envelope offset, the beats of the clock lasers with the comb are chosen to have the same sign. In addition, these beat frequencies are chosen to be identical ($f_{\mathrm{b}\uparrow} = f_{\mathrm{b}\downarrow} = 20$ MHz), phase locked to RF references having a fixed phase relation.[15] The FNC setups are also intentionally identical for both legs; i.e., the FNC beat notes are phase locked to the same RF reference derived from a single output of a DDS, and AOM2 and AOM3 diffract the same order ($\{+1\}$) so that $f_{A2} = f_{A3} = 80$ MHz.

The "raw" or perturbed clock frequency not corrected for systematics is

$$f_{\mathrm{clock}}^{\mathrm{raw}} = f_\downarrow^{\mathrm{mol}} - f_\uparrow^{\mathrm{mol}} = (N_\downarrow - N_\uparrow) f_{\mathrm{RR}} - f_{\mathrm{A1},\mathbf{x}}, \tag{5.34}$$

where we add a subscript to explicitly state that $f_{\mathrm{A1},\mathbf{x}}$ is the AOM1 frequency corresponding to the line center of the fitted clock spectrum taken under the experimental parameters $\mathbf{x}$. This is to be contrasted with the *unperturbed* clock frequency, where we extrapolate to the idealized situation when all external experimental parameter values are zero ($\mathbf{x} = 0$),

$$f_{\mathrm{clock}} = f_\downarrow^{\mathrm{mol}} - f_\uparrow^{\mathrm{mol}} + f_{\mathrm{corr.}} = (N_\downarrow - N_\uparrow) f_{\mathrm{RR}} - f_{\mathrm{A1},\mathbf{x}_{\mathrm{opt}}} + f_{\mathrm{corr.}}. \tag{5.35}$$

Here, $f_{\mathrm{corr.}}$ is the total frequency correction due to systematic shifts [Table 5.3] for the operational experimental parameter values, $\mathbf{x}_{\mathrm{opt}}$. That is,

$$f_{\mathrm{corr.}} \equiv f_{\mathrm{A1},\mathbf{x}_{\mathrm{opt}}} - f_{\mathrm{A1},0}. \tag{5.36}$$

Finally, the "clock shift" is defined to be the frequency shift of the perturbed clock frequency relative to the unperturbed clock frequency,

$$\Delta f_{\mathrm{clock}} \equiv f_{\mathrm{clock}}^{\mathrm{raw}} - f_{\mathrm{clock}} = f_{\mathrm{A1},0} - f_{\mathrm{A1},\mathbf{x}}. \tag{5.37}$$

As f_{RR} is tunable over a wide range, it can be judiciously chosen such that f_{A1} is close to the specified center frequency of AOM1 (80 MHz) to achieve good diffraction efficiency.[16] The molecular clock resonances occur at $f_{A1} \approx 82.6$ MHz

[15] This is guaranteed by having all synthesizers referenced to a single lab timebase.

[16] Note that because the FNC beats occur at $2 \times f_{\mathrm{A2,A3}}$, as a precaution, f_{A1} should also be made sufficiently different from f_{A2} to avoid the situation where leakage higher diffraction orders from AOM1 and subsequent multi-path reflections interfere with the FNC setup for the upleg. Here, we fortuitously avoided this by a $\approx$2.6 MHz difference.

if we choose $f_{RR} \approx 250035478.54$ Hz, $N_\uparrow = 1510825$, $N_\downarrow = 1638108$. The clock lasers are $10 \times$ FSR $\approx$ 14.973 GHz blue detuned from the transition between the $(1)0_u^+(11, 1)$ intermediate state and their respective ground states, where FSR is the free spectral range of reference cavity (see Sect. 3.1.1).

5.5 Systematic Evaluation, and General Methodology

Table 5.3 details the uncertainty budget of the molecular clock under the operational conditions of this work. Summing the uncertainties of all contributors in quadrature, we report a total systematic uncertainty of 4.6×10^{-14}. We leverage the short-term frequency stability of our reference cavity to average down the uncertainty of a given systematic.[17] We probe the clock transition in an interleaved fashion; i.e., we alternate an experimental parameter between two values and record f_{A1} for the corresponding pair of line centers. This is repeated to gather statistics.

Note that for the density shift measurements, the molecule numbers were changed cycle-to-cycle (i.e., interlaced). For the probe and lattice light shift measurements, the laser intensities were changed scan-to-scan, due to the speed at which the motorized neutral density filters could reliably switch positions.

To subdue the effect of residual cavity drift not canceled by the feedforward acousto-optic modulator, we employ three-point string analysis [36]. Let $\mathbf{y} = \{y_1, y_2, \ldots, y_n\}$ denote the series of AOM1 frequencies, f_{A1}, corresponding to the clock line centers when a single experimental parameter is repeatedly varied between a reference value (x_{ref}) and a different predetermined value ($x = x_{ref} + \Delta x$).

Table 5.3 Systematic uncertainty budget for the strontium molecular clock under operating conditions. Listed are the frequency corrections, $f_{corr.}$, and their uncertainties. The signs of the corrections are defined such that these values add to the perturbed clock frequency to give the unperturbed clock frequency, f_{clock}. All values are expressed in fractional units ($\times 10^{-14}$)

Systematic	Correction	Uncertainty
Lattice stark ($E1$, $M1$, $E2$)	100.1	3.4
Lattice stark (hyperpolarizability)	−50.8	1.9
Probe stark (total)	31.5	2.2
BBR	−2.2	0.4
Density	−0.6	0.3
Quadratic zeeman	0	0.05
dc Stark	0	< 0.1
Doppler and phase chirps	0	< 1
Lattice tunneling	0	< 0.1
Line pulling	0	< 0.1
Scan-and-fit	0	< 0.6
Total	77.9	4.6

[17] The interleaved shift measurements are performed quickly (~20 s for a single scan of the clock spectrum), during which f_{RR} does not change by a meaningful amount. Contamination by such drifts is also suppressed by point-string analysis. Nevertheless, we have quantified potential systematic errors due to the reference cavity under the designation "Scan-and-fit".

Here, n is an even number because we always record pairs of line centers in a series of interleaved measurements. Ideally, Δx should be large enough to induce a measurable frequency shift. We calculate the frequency shift between consecutive points as

$$\Delta y_1 = \frac{y_1 + y_3}{2} - y_2,$$
$$\Delta y_2 = y_3 - \frac{y_2 + y_4}{2},$$
$$\Delta y_3 = \frac{y_3 + y_5}{2} - y_4,$$
$$\vdots$$
$$\Delta y_{n-2} = y_{n-1} - \frac{y_{n-2} + y_n}{2}.$$

Essentially, we assume that the cavity drift is linear in the duration between points y_i and y_{i+2}. If the cycle times are equal,[18] then $(y_i + y_{i+2})/2$ estimates the line center as if it were measured at the same time as y_{i+1}. A visual representation is shown in Fig. 5.11. Thus, the cavity drift should be greatly nulled using the differences Δy_i. The "frequency shift", $\Delta f_{\mathrm{A1}} = f_{\mathrm{A1},x} - f_{\mathrm{A1},x_{\mathrm{ref}}}$, is calculated as the weighted average of these differences. A drawback of point string analysis is that we have artificially correlated neighboring elements and doubled the number of frequency shifts extracted from the dataset—the statistically independent, non-overlapping, set of frequency shifts $\{y_1 - y_2, y_3 - y_4, \ldots\}$ only has $n/2$ elements. Hence, we further multiply the revised standard error of the weighted mean obtained using point string analysis, $\tilde{\sigma}_{\Delta \mathrm{f}_{\mathrm{A1}}}$, by a numerical factor of approximately $\sqrt{2}$ to extrapolate back to the expected level of statistics had we used independent, non-overlapping, successive pairs of interleaved measurements to extract the frequency shifts.

After collecting Δf_{A1} and their uncertainties for various Δx, we perform weighted fitting of a model function $f^{\mathrm{ref}}_{\mathrm{model}}(x - x_{\mathrm{ref}})$. We quote and utilize the revised standard errors (i.e., scaled up by $\sqrt{\chi^2_{\mathrm{red}}}$) of the fitted parameters of all weighted polynomial fits. Logically, the frequency shift is identically zero if the parameter is unchanged ($\Delta x = 0$). We demonstrate the extent to which this is valid by repeatedly scanning the molecular line while keeping the experimental parameters and conditions constant to the best of our ability. We observe that the Allan deviation [Fig. 5.12] of the frequency differences extracted using three-point string analysis averages to $\approx 1 \times 10^{-14}$, with a weighted mean consistent with

[18] For the density shift evaluations, the cycle times ($\Delta t_1 = \Delta t_3 > \Delta t_2, \ldots$) can be rather dissimilar due to sample preparation. We have experimented with assigning "weights" to the measurement points; i.e., $\Delta y_1 = (\Delta t_1 y_1 + \Delta t_2 y_3)/(\Delta t_1 + \Delta t_2) - y_2$, $\Delta y_2 = y_3 - (\Delta t_2 y_2 + \Delta t_1 y_4)/(\Delta t_1 + \Delta t_2), \ldots$. However, we find that the calculated clock shifts are not significantly different from those using simple three-point strings ($\Delta t_1 = \Delta t_2, \ldots$).

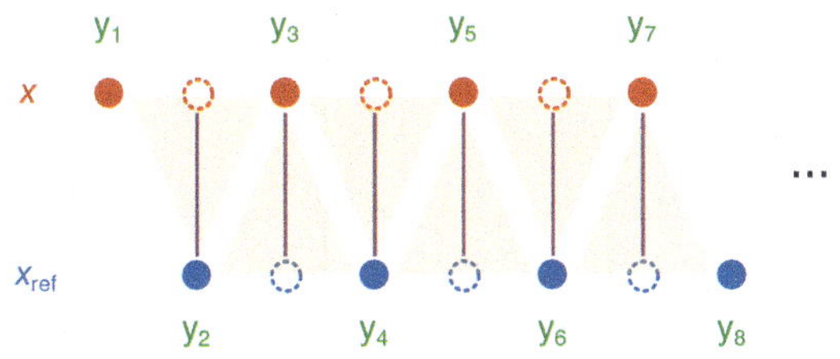

Fig. 5.11 Pedagogical illustration of three-point strings used to average the corresponding frequency shift when an experimental parameter is repeatedly changed between x_{ref} and x, color-coded blue and red respectively. This approach suppresses contamination of the computed frequency differences by slow ambient drifts, such as (but not limited to) reference cavity drift. The line centers y_i (solid red and blue circles, labeled by green letters) are obtained from fitting Lorentzians over the scanned molecular resonances. We estimate the instantaneous drift using adjacent points taken under the same experimental parameters (y_i and y_{i+2}), and use it to predict the expected line center (dotted red and blue circles) as if it were measured at the same time as y_{i+1}. The frequency shift is computed as the weighted average of these simulated simultaneous differences (represented by the vertical purple lines). Lastly, we scale up the revised standard error of the mean by a factor $\sim \sqrt{2}$ because the number of three-point string differences $(n-2)$ is approximately $2\times$ larger than the number of statistically independent differences $(n/2)$

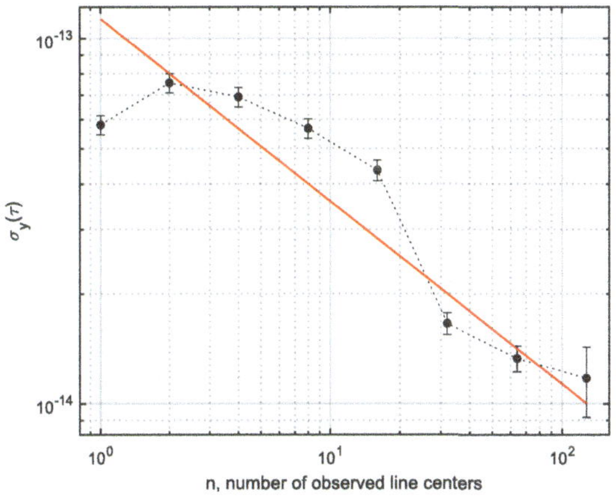

Fig. 5.12 Overlapping Allan deviation of three-point string frequency differences for the situation when the experimental parameters are left unchanged. The Allan deviation averages down as $\sim 10^{-13}/\sqrt{n}$, where n is the number of observed line centers. Put another way, in order to resolve a frequency shift at the 10^{-14} level, we will have to average $\sim 10^2$ clock lines. The situation should improve with longer lifetimes and coherence times (to resolve smaller linewidths), and better reference cavity stability

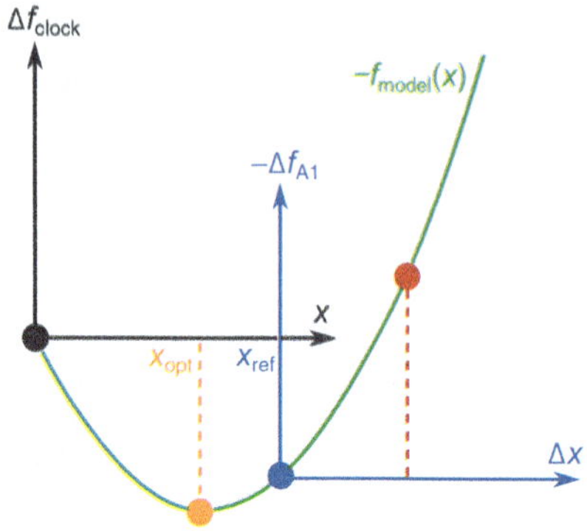

Fig. 5.13 Graphical relationship between the clock shift (Δf_{clock}), the AOM1 frequency shift (Δf_{A1}), and the functional model for frequency extrapolation (f_{model}), plotted versus the value of a given experimental parameter (x). See main text for details

zero. Therefore, the model has a constraint that $f^{\text{ref}}_{\text{model}}(0) = 0$. The extrapolated frequency correction for the clock at the operational parameter value is

$$f_{\text{corr.}} = f^{\text{ref}}_{\text{model}}(x_{\text{opt}} - x_{\text{ref}}) - f^{\text{ref}}_{\text{model}}(-x_{\text{ref}}). \tag{5.38}$$

Equation (5.38) follows from a straightforward exercise in coordinate transformation [Fig. 5.13]. For simplicity of discussion, suppose that the model perfectly fits the frequency shifts (i.e., $\chi^2 = 0$) so that we may use equal signs for the following few equations. We start with

$$f^{\text{ref}}_{\text{model}}(x - x_{\text{ref}}) = f_{\text{A1},x} - f_{\text{A1},x_{\text{ref}}}.$$

Setting $x = 0$, we get

$$f^{\text{ref}}_{\text{model}}(-x_{\text{ref}}) = f_{\text{A1},0} - f_{\text{A1},x_{\text{ref}}},$$

which incidentally is the frequency correction for the clock at the reference parameter x_{ref}. We seek the relationship between $f^{\text{ref}}_{\text{model}}(x - x_{\text{ref}})$ and the model function centered at $x = 0$,

$$\begin{aligned} f_{\text{model}}(x) &= f_{\text{A1},x} - f_{\text{A1},0}, \\ &= -\Delta f_{\text{clock}}, \end{aligned}$$

where the second line is true by definition [Eq. (5.37)]. In other words,

$$f^{\text{ref}}_{\text{model}}(x - x_{\text{ref}}) = f_{\text{model}}(x) - f_{\text{model}}(x_{\text{ref}}). \tag{5.39}$$

Next, we eliminate $f_{\text{A1},x_{\text{ref}}}$ from our expressions,

$$f_{\text{model}}(x) = \left[f^{\text{ref}}_{\text{model}}(x - x_{\text{ref}}) + f_{\text{A1},x_{\text{ref}}}\right] - \left[f^{\text{ref}}_{\text{model}}(-x_{\text{ref}}) + f_{\text{A1},x_{\text{ref}}}\right],$$

$$= f^{\text{ref}}_{\text{model}}(x - x_{\text{ref}}) - f^{\text{ref}}_{\text{model}}(-x_{\text{ref}}). \tag{5.40}$$

As a sanity check, we see that indeed $f_{\text{model}}(0) = 0$; i.e., the total clock shift of the unperturbed clock is zero. To arrive at Eq. (5.38), we recall from Eq. (5.36) that the frequency correction for the clock at the *operational* parameter value is[19]

$$f_{\text{corr.}} = f_{\text{model}}(x_{\text{opt}}).$$

5.6 Sources of Systematic Error

5.6.1 Lattice Light Shift

As discussed in Sect. 5.3, magic trapping conditions can be engineered for this pair of vibrational clock states by off-resonantly addressing $X(0,0) \rightarrow (1)1_u(9,1)$ with the lattice ($\Delta_m = 4.494(1)$ GHz). Importantly, the neighboring $(1)1_u(v',1)$ rovibronic resonances are spaced at intervals of ~2 THz, and may cause deleterious shifts due to lattice light impurity (e.g., amplified spontaneous emission (ASE) [37]). To mitigate this, the lattice light derives from a Ti:sapphire laser. The lattice light that is delivered to the molecules is phase-stabilized to the optical frequency comb[20] by actuating on a voltage-controlled oscillator (VCO) that drives an external AOM to correct for fast phase excursions. In general, acousto-optic modulators have a limited bandwidth; i.e., the diffraction efficiency falls quickly outside a finite range of RF frequencies. To avoid (or at least slow down) such a scenario, the servo output monitor of the main servo controller (that tunes the VCO) is sent to a second servo controller as an error signal input. The output of this second controller actuates on the "fast" resonantor piezo[21] of the Ti:sapphire laser, and the setpoint is tuned to keep the output frequency of the VCO close to the specified center frequency of the AOM (where the diffraction is optimal). The bandwidth of this second controller is set to be small to avoid cross-talk with the main controller. While this external AOM-based phase locking setup is sufficient for the work in this thesis, precious laser power is wasted in the zeroth AOM order, and the lock is rather delicate. Future work requiring long-term operation of the molecular clock may benefit from injection locking a Ti:sapphire laser with a diode laser robustly phase-locked to the comb.

[19] Note that the relative signs will be different if the downleg is scanned instead of the upleg.

[20] This also permits the lattice frequency, $f_{\text{latt}} = c/\lambda_{\text{latt}}$, to be determined with kHz-level accuracy. Note that the M-NIR port of the frequency comb was used for lattice frequency stabilization, whose zeroth comb tooth is at f_{CEO}; see Sect. 3.1.2.

[21] We find that the "slow" piezo has an overly sensitive transfer function; i.e., small voltage noise thwarts proper engagement of the phase lock.

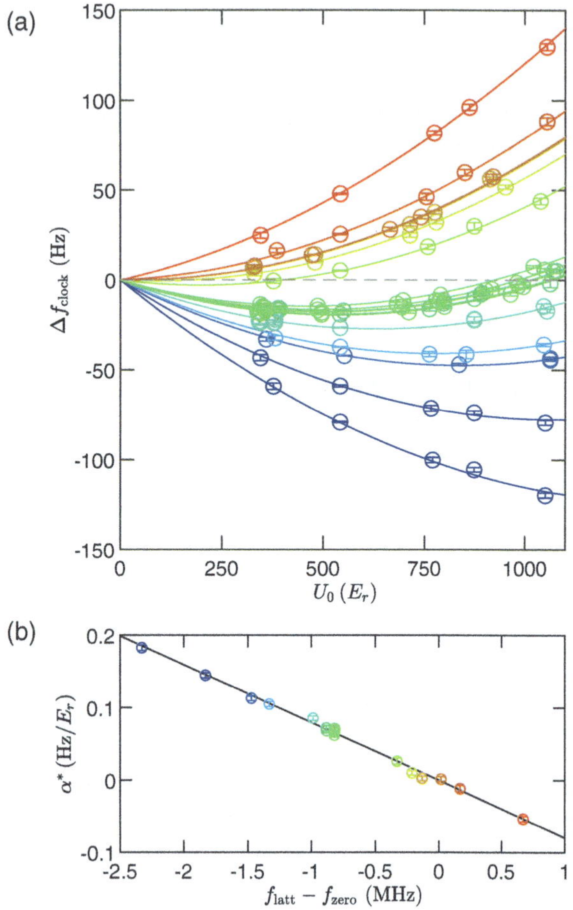

Fig. 5.14 Clock shifts due to the lattice light. (**a**) Non-linear shifts of the molecular clock frequency versus trap depth. For a given lattice frequency (color coded), we make interleaved measurements of clock shifts (open circles) with respect to a reference trap depth ($\sim$500 E_{r}), and fit the data to parabolas (solid lines) with a global quadratic parameter, $-\beta^*$. (**b**) Linear light shift coefficient, α^*, versus lattice frequency (color code matches (**a**)), and the linear fit (black solid line). α^* is predominantly due to the $E1$ differential polarizability and is nulled at f_{zero}. By tuning α^*, we can find conditions where the sensitivity of Δf_{clock} to fluctuations in U_0 is minimal at our operational trap depth of 487(4) E_{r} (dark green points). Error bars represent 1σ uncertainties

To further suppress ASE impurity at the magic detuning, the light is filtered through a linear cavity (finesse of 50, and free spectral range of 2.9 GHz) before delivery to the experiment by a single-mode polarization maintaining fiber. During a normal clock run under operational conditions, a directionally stable, weak reflection from the vacuum window is used for lattice intensity stabilization. The linear lattice polarization defines the quantization axis for the magnetically insensitive $X^1\Sigma_g^+$ states [26].

We investigate the effect of lattice light over a range of f_{latt}. At each f_{latt} we make interleaved measurements of the clock shifts, alternating the trap depth U_0 between a reference depth and four other depths spanning from 300 E_{r} to 1100 E_{r}. This is accomplished by the use of a motorized neutral density filter placed before the lattice is injected into its optical fiber.[22] The trap depths are determined from the axial trapping frequencies [Eq. 2.32]. The axial (f_{ax}) and radial (f_{rad}) trap frequencies for the molecules are measured with resolved-sideband spectroscopy at the operational magic lattice wavelength [Fig. 2.7a, b]. The $1/e^2$ beam waist (w_0) of the lattice at its focus is related to the trap frequencies; i.e., $f_{ax}^2/f_{rad}^2 = w_0^2(2\pi/\lambda_{latt})^2/2$. In this work, $w_0 = 36(1)$ μm.

As shown in Fig. 5.14a, our measurements reveal non-linear light shifts as a consequence of molecular hyperpolarizability. Small corrections ($< 0.3 \times 10^{-14} \times f_{clock}$) were made to account for density shifts.[23] The quadratic lattice scattering rates discussed in Sect. 5.3.3 may share the same origin as these non-linear shifts.

In order to characterize the lattice light shifts, we adopt the thermal model described in Ref. [38] and write the clock shifts as

$$\Delta f_{clock} = -\alpha^* U_0 - \beta^* U_0^2, \tag{5.41}$$

where α^* and β^* are empirically obtained from parabolic fits to the measured differential shifts.[24] These parameters are effective values dependent on the trapping conditions: α^* is related to the differential electric-dipole ($E1$), magnetic-dipole ($M1$) and electric-quadrupole ($E2$) polarizabilities, while β^* is related to the differential hyperpolarizability. Further justification of Eq. (5.41) can be found in Refs. [38, 39] and the appendices therein. Crucially, the polynomial form of Eq. (5.41) hinges on a linear scaling of the sample temperature with U_0, which we verify to hold true for our molecules using Raman carrier thermometry [Fig. 5.15]. Given our trap frequencies, sample temperatures, and the linear scaling of sample

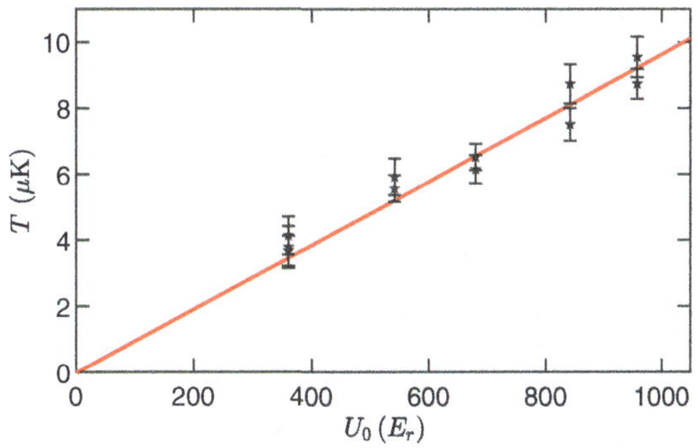

Fig. 5.15 Measurement of the molecular temperature as a function of trap depth (black stars) using Raman carrier thermometry. Error bars denote 1σ uncertainties. A linear fit (solid red line) with the intercept fixed at the origin describes the data well ($\chi^2_{red} = 0.92$)

[22] The passive stability of lattice light is at the 1% level. To ease experimental complexity, the lattice intensities were not actively stabilized for the lattice shift measurements.

[23] Accounting for the density shift changes the fitted β^* value by approximately 0.1×10^{-5} Hz$/E_r^2$. To be conservative, we add this in quadrature with the uncertainty of β^* from the fitting.

[24] The parabola fits are constrained such that $f^{ref}_{model}(0) = 0$, which we have verified to be true in the present experimental setup at the level of $< 10^{-14}$; see Fig. 5.12.

temperature with trap depth, the polynomial fit is a reasonable approximation, and the $M1$-$E2$ shifts (that microscopically scale as $\sqrt{U_0}$ [19]) are included in the α^* effective parameter. In future work, calculating the differential $M1$ and $E2$ polarizabilities would help quantify the error associated with the thermal model.

The fits give $\beta^* = -6.81(22) \times 10^{-5}\,\text{Hz}/E_r^2$ as a global parameter. Additionally, the results for α^* versus f_{latt} are shown in Fig. 5.14b, and a linear fit yields a sensitivity slope $\partial\alpha^*/\partial f_{\text{latt}} = -0.0796(16)\,\text{Hz}/(\text{MHz}\,E_r)$ as well as an x-intercept $f_{\text{zero}} = 298\,368\,568.844(21)\,\text{MHz}$. Operating the molecular clock at a trap depth of $U_{\text{opt}} = 487(4)\,E_r$ and $f_{\text{latt}} - f_{\text{zero}} = -0.821(21)\,\text{MHz}$, we determine the correction terms to be $\alpha^* U_{\text{opt}} = 31.8(1.1)\,\text{Hz}$ and $\beta^* U_{\text{opt}}^2 = -16.2(6)\,\text{Hz}$, summing to a fractional correction of $49.3(3.8) \times 10^{-14}$. Under these conditions, Δf_{clock} is first-order insensitive to changes in U_0 (dark green points in Fig. 5.14).

To test if higher-order polynomial terms are statistically significant, we fit the lattice light shifts to a cubic polynomial, with the quadratic and cubic coefficients as global fit parameters. The data suggests a cubic coefficient of $-1.2(9) \times 10^{-8}\,\text{Hz}/E_r^3$. While the addition of a cubic term shifts the value of f_{zero}, the estimated frequency correction, in this case, remains consistent with the applied correction within their uncertainties. As such, we limit our fitting to a quadratic polynomial for the present evaluation at the 10^{-14} level.

While we leave a full theoretical investigation the observed molecular hyperpolarizability to future work, it is possible to gain some insights by considering a toy model [13] involving three states;[25] i.e., $X(0,0)$, $X(0,2)$, and $(1)1_u(9,1)$. Limiting our analysis to the simplest, idealized case with linear polarization, straightforward application of formulas in Ref. [4] predicts, for the absolute ground state, that scalar component of the hyperpolarizability is

$$\begin{aligned}\gamma^{\text{pred.}} = &-\frac{4}{3}\begin{Bmatrix}1&1&0\\0&0&1\end{Bmatrix}\begin{Bmatrix}1&1&0\\0&0&1\end{Bmatrix}\begin{Bmatrix}0&0&0\\0&0&0\end{Bmatrix}\frac{|\langle f||d^{(1)}||i\rangle|^4}{\hbar^3}\frac{4\omega_{fi}(\omega_{fi}^2+\omega^2)}{(\omega_{fi}^2-\omega^2)^3}\\ &+\frac{8\sqrt{5}}{3}\begin{Bmatrix}1&1&2\\0&2&1\end{Bmatrix}\begin{Bmatrix}1&1&2\\0&2&1\end{Bmatrix}\begin{Bmatrix}2&2&0\\0&0&2\end{Bmatrix}\frac{|\langle f||d^{(1)}||i\rangle|^4}{\hbar^3}\frac{4\omega_{fi}^2}{(\omega_{fi}^2-\omega^2)^2\Delta\omega_{\text{rot}}},\end{aligned}$$

where $\omega = 2\pi f_{\text{latt}}$ is the angular frequency of the lattice laser, $\Delta\omega_{\text{rot}} = 2\pi \times 3.146(1)\,\text{GHz}$ is the rotational splitting between $X(0,0)$ and $X(0,2)$ in angular frequency units [Table 3.4], and $\omega_{fi} = 2\pi \times 298\,364\,075(1)\,\text{MHz}$ is the *resonant* angular transition frequency of $X(0,0) \rightarrow (1)1_u(9,1)$ [Table 3.6]. Following the nomenclature of Sects. 5.1.2 and 5.1.3, the state labels $|i\rangle$ and $|f\rangle$ here are shorthand for $\{v_i = 0, J_i = 0, M_i = 0, \Omega_i = 0, n_i = X^1\Sigma_g^+\}$, and $\{v_f = 9, J_f = 1, M_f = 0, \Omega_f = 1, n_f = (1)1_u\}$ respectively. We ignore centrifugal effects, such that the *reduced* matrix element $|\langle f||d^{(1)}||i\rangle|$ has the same value for transitions between

[25] Using Fig. 3.7 as a stand-in illustration, the level structure of these three states would look like $X(0,2) \rightarrow |1\rangle$, $X(0,0) \rightarrow |2\rangle$ and $(1)1_u(9,1) \rightarrow |4\rangle$.

$X(0, J = 0, 2)$ and $(1)1_u(9, 1)$. As before, the array with curly brackets is the Wigner 6-j symbol. Assuming that $X(62, 0)$ is unperturbed, this hyperpolarizability coefficient would contribute a clock shift

$$\Delta f_{\text{clock}}^{\text{pred.}} = -\left(-\frac{1}{64}\gamma^{\text{pred.}}\mathcal{E}_0^4\right)/h,$$
$$= \frac{9|\Omega_R|^4}{2\pi}\left[-\frac{1}{36}\frac{\omega_{fi}(\omega_{fi}^2+\omega^2)}{(\omega_{fi}^2-\omega^2)^3}+\frac{1}{90}\frac{\omega_{fi}^2}{(\omega_{fi}^2-\omega^2)^2\Delta\omega_{\text{rot}}}\right], \quad (5.42)$$

where $|\Omega_R| = |\langle f|d^{(1)}|i\rangle|\mathcal{E}_0/\hbar = \left(1/\sqrt{3}\right)|\langle f||d^{(1)}||i\rangle|\mathcal{E}_0/\hbar$ is the absolute value of the angular Rabi frequency between $X(0, 0)$ and $(1)1_u(9, 1)$ induced by the lattice laser light [Eqs. (4.2) and (5.11)]. To make contact with experiment, we utilize Eq. (5.19) to rewrite Eq. (5.42) as

$$\Delta f_{\text{clock}}^{\text{pred.}} = \frac{\hbar^2\Delta_m^2\omega^4}{\pi M^2c^4}\left[-\frac{1}{2}\frac{\omega_{fi}(\omega_{fi}^2+\omega^2)}{(\omega_{fi}^2-\omega^2)^3}+\frac{1}{5}\frac{\omega_{fi}^2}{(\omega_{fi}^2-\omega^2)^2\Delta\omega_{\text{rot}}}\right]$$
$$\left(\frac{\alpha_0^{\text{b.g.}}}{\alpha_{62}}-1\right)^2\frac{U_0^2}{E_{\text{r}}^2}, \quad (5.43)$$

which makes it explicit that this shift scales quadratically with trap depth U_0. Recall that the baseline polarizability ratio $\alpha_0^{\text{b.g.}}/\alpha_{62} = 1.404(7)$ around ~1005 nm, and the magic detuning for the present pair of clock states and chosen $X \to 1_u$ resonance is $\Delta_m = 2\pi \times 4.494(1)$ GHz (see Sect. 5.3, in particular Figs. 5.1 and 5.3, and Table 5.1). Thus, the parameters that enter Eq. (5.43) are all experimentally determined using accurate, frequency-based measurements. Evaluating Eq. (5.43) at $\omega = 2\pi f_{\text{zero}}$, and making the comparison with Eq. (5.41), we predict the quadratic coefficient to be $\beta^{\text{pred.}} = -7 \times 10^{-5}\ \text{Hz}/E_{\text{r}}^2$, whose sign and order of magnitude[26] both correspond well with the experimentally measured β^*. It is also worth noting that the measured sensitivity slope $\partial\alpha^*/\partial f_{\text{latt}}$ compares well with the rough estimate of $-0.1\ \text{Hz}/(\text{MHz}\ E_{\text{r}})$ using

$$-\frac{|\Omega_R|^2}{4\Delta_m^2} \simeq -\left(\frac{\alpha_0^{\text{b.g.}}}{\alpha_{62}}-1\right)\frac{\hbar\omega^2}{2Mc^2\Delta_m}\frac{U_0}{E_{\text{r}}}. \quad (5.44)$$

[26] Within the square brackets of Eq. (5.43), the first term can be interpreted as the higher-order correction in the coupling of $X(0, 0)$ with $(1)1_u(9, 1)$, while the second term accounts for the lattice light-mediated Raman-type coupling between $X(0, 0)$ and $X(0, 2)$ via $(1)1_u(9, 1)$. The former is approximately twice as large as the latter, and the values of both terms possess the same sign.

A caveat of Eqs. (5.43) and (5.44) is the assumption that the molecules sample the peak intensity at each microtrap—an idealization that should be modified to account for $T \neq 0$ [Fig. 5.15].

5.6.2 Probe Light Shift

Probe light shifts pose an inherent challenge for two-photon spectroscopy. This is even more so for scalar clock states ($J = 0$) that preclude the use of laser polarization-based cancellation schemes [40]. Here, the clock shifts scale linearly as the probe intensities are low, and are related to the differential polarizability at the respective probe wavelength (λ_p),

$$\Delta f_{\text{clock}} = \frac{I_p}{2h\epsilon_0 c}\left[\alpha_0(\lambda_p) - \alpha_{62}(\lambda_p)\right], \tag{5.45}$$

where α_v is the $E1$ polarizability for the vibrational state v, I_p is the probe laser intensity, and here $p \in \{\uparrow, \downarrow\}$ specifies the laser: upleg ($\uparrow$) or downleg ($\downarrow$). Figure 5.16 shows that linear extrapolation of probe shifts suffices for a molecular clock at the few 10^{-14} level.

While tailored pulse sequences to alleviate probe light shifts have been proposed [41–44], for this evaluation we opted for a more straightforward strategy. We can minimize the total probe light shift by using so-called balanced intensity ratios satisfying the condition $I_\uparrow\left[\alpha_0(\lambda_\uparrow) - \alpha_{62}(\lambda_\uparrow)\right] = -I_\downarrow\left[\alpha_0(\lambda_\downarrow) - \alpha_{62}(\lambda_\downarrow)\right]$. At the same time, a large Raman detuning—relative to the intermediate $(1)0_u^+(11, 1)$ excited state—is preferred so that off-resonant scattering from the probes have a negligible effect on the accessible coherence times. Figure 5.16 demonstrates that such conditions exist in our clock for blue detunings where the baseline polarizability differences at the probe wavelengths have opposite signs, in agreement with our polarizability model (see Sect. 5.1.3). We operate at a Raman detuning of +14.973

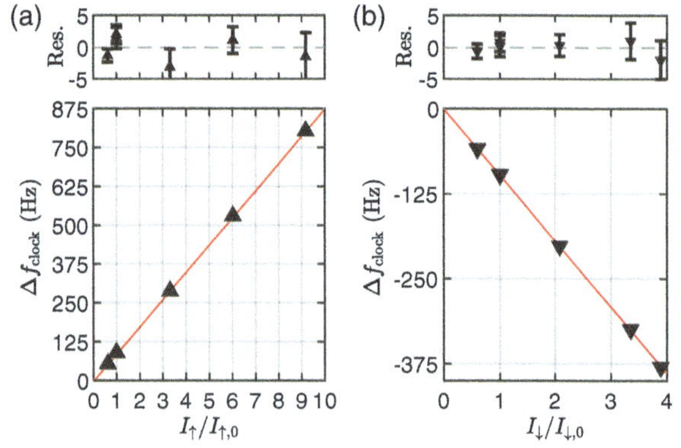

Fig. 5.16 Clock shifts at the operational Raman detuning as a function of (**a**) the upleg laser intensity, and (**b**) the downleg laser intensity. The horizontal axes are normalized by the respective operational intensities, $I_{\uparrow,0}$ and $I_{\downarrow,0}$. Solid lines are linear fits to the data. Residuals are plotted in units of Hz. Error bars represent 1σ uncertainties

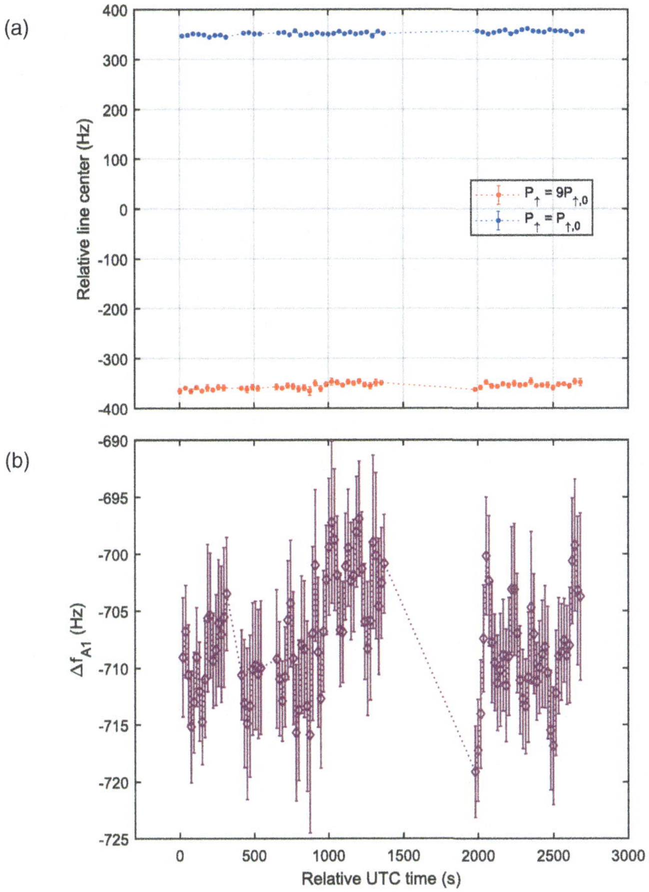

Fig. 5.17 (**a**) A representative time series of measured line centers during a systematic evaluation. An interleaved measurement alternates between two controlled values of an experimental parameter (here, it is the upleg power, $P_\uparrow$). (**b**) The frequency shifts (purple diamonds) are calculated using three-point string analysis. We take the weighted mean of these differential shifts, and scale up the revised standard error (i.e., the standard error inflated by $\sqrt{\chi^2_{\text{red}}}$) by a statistical factor of approximately $\sqrt{2}$

GHz, which is over 3 orders of magnitude greater than the 5 MHz natural linewidth of the intermediate state [Fig. 4.4].

We evaluate Δf_{clock} for each leg separately. Using a motorized neutral density filter, we switch between two intensity values for one leg while keeping that of the other leg constant at its operational value [Fig. 5.17]. The π-pulse durations are adjusted accordingly. Typical settings for the interleaved measurements are $(P_{\uparrow,0}, 9P_{\uparrow,0})$, and $(P_{\downarrow,0}, 3.5P_{\downarrow,0})$ where $P_{p,0} = I_{p,0}(\pi w_p^2/2)$ are the operational powers measured with a calibrated power meter immediately before the vacuum window. These shifts are scaled by the measurement lever arms to obtain the clock corrections at the operational settings: $-(\Delta f_{\text{clock}}/\Delta P_p) \times P_{p,0}$. We find the corrections to be $-277.5(1.4) \times 10^{-14}$ for the upleg, and $309.0(1.7) \times 10^{-14}$ for the downleg. The uncertainties in probe intensities have been propagated into those of the quoted frequency corrections. Drifts in ΔP_p are at the sub-percent ($\sim$0.5%) level over the $\sim$2000 s duration for each probe light shift evaluation, and the weighted averages typically have $\chi^2_{\text{red}} \sim 1$.

We observe the beam profiles of the clock lasers on a camera and estimate the $1/e^2$ beam waists of the upleg and downleg at the molecules to be 89(5) and 114(20) μm, respectively. The uncertainties are large due to the technical difficulty in determining beam waists accurately. However, accurate knowledge of the beam waists w_p is not necessary as they are robust during an evaluation, and are common factors that drop out in calculations. Long-term intensity drifts (e.g., due to instabilities in beam pointing instability or time-varying servo bumps) may be monitored and countered by benchmarking the probe intensities using the molecules (e.g., through an Autler-Townes splitting, an on-resonance scattering rate, or the two-photon Rabi oscillation frequency), which we leave to future work.

5.6.3 *Blackbody Radiation Shift*

Homonuclear dimers are infrared inactive, conferring natural immunity to blackbody radiation (BBR). According to formulas derived in Ref. [45], the level shift due to BBR (in atomic units) is approximately

$$\Delta_v^{\text{BBR,au}} \approx -\frac{2}{15}(\alpha_{\text{fs}}\pi)^3 (T^{\text{eff,au}})^4\, \alpha_v^{\text{au}}(0)(1+\eta_{\text{dyn},v}). \tag{5.46}$$

Here, the superscript "au" implies the value of a quantity in atomic units. α_{fs} is the fine-structure constant, and $\eta_{\text{dyn},v}$ is the so-called dynamic term

$$\eta_{\text{dyn},v} = \frac{(80/63)\pi^2}{\alpha_v^{\text{au}}(0)}(T^{\text{eff,au}})^2 \times \sum_f \frac{|\langle f||d^{\text{au}}||i\rangle|^2}{(\omega_{fi}^{\text{au}})^3} \left[1 + \frac{21\pi^2(T^{\text{eff,au}})^2}{5(\omega_{fi}^{\text{au}})^2} + \frac{336\pi^4(T^{\text{eff,au}})^4}{11(\omega_{fi}^{\text{au}})^4}\right]. \tag{5.47}$$

In Eqs. (5.46) and (5.47), the numerical values of the electron mass, elementary charge, and reduced Planck constant are set to unity. State notation is consistent with Eq. (5.10). The clock shift is

$$\Delta f_{\text{clock}} = \Delta_{62}^{\text{BBR}} - \Delta_0^{\text{BBR}}. \tag{5.48}$$

Following Ref. [46], the effective temperature (T^{eff}) that enters into the BBR shift calculation is such that

$$(T^{\text{eff}})^4 = \sum_i \frac{\Omega^{\text{eff}}_{\text{angle},i}}{4\pi} T_i^4, \tag{5.49}$$

where T_i is the temperature of the i-th surface. Short of finite-element modeling, we may make a basic estimate for the effective solid angle ($\Omega^{\text{eff}}_{\text{angle},i}$) subtended by the i-th surface surrounding the molecular sample as

$$\frac{\Omega^{\text{eff}}_{\text{angle},i}}{4\pi} = \frac{\Omega_{\text{angle},i}\,\eta_i}{\sum_i \Omega_{\text{angle},i}\,\eta_i}, \tag{5.50}$$

where η_i is the surface emissivity and $\Omega_{\text{angle},i}$ is the geometric solid angle. The total effective solid angle is normalized such that $\sum_i \Omega^{\text{eff}}_{\text{angle},i} = 4\pi$. We use values for the emissivity of various materials from available literature [47–51]. These are 0.91 for glass (fused silica), 0.54 for sapphire, and 0.08 for stainless steel.

The sapphire window facing the Zeeman slower is held within a nipple heated to 430(10) K, and subtends a geometric solid angle of 0.04 sr. Among the fused silica window viewports with direct line-of-sight to the molecules, there are 8 with a diameter of 33.78 mm, 4 with a diameter of 69.85 mm, and 6 with a diameter of 114.3 mm. The diameter of the spherical vacuum chamber is approximately 240(10) mm, and the surface area consisting of stainless steel is approximated as the spherical surface area subtracted by the total area encompassed by the viewports.

At the present level of precision, it is enough to estimate the temperature environment of the stainless steel vacuum chamber using four negative temperature coefficient (NTC) thermistors affixed to its exterior. The largest (smallest) sensor reading is $T_{\text{c,max}}$ ($T_{\text{c,min}}$). We model the temperature gradient as a rectangular distribution [52] and estimate the temperature of the vacuum chamber to be $(T_{\text{c,max}} + T_{\text{c,min}})/2 = 302$ K, with an uncertainty of $(T_{\text{c,max}} - T_{\text{c,min}})/\sqrt{12} = 1$ K. Conservatively, the fused silica windows are within ± 2 K of the temperature of the stainless steel chamber. The line of sight from the molecules to the hot oven is blocked using an in-vacuum mechanical shutter during clock spectroscopy.

We estimate an effective blackbody temperature $T^{\text{eff}} = 303(5)$ K, which implies a frequency correction ($-\Delta f_{\text{clock}}$) of $-0.70(14)$ Hz. The uncertainty is dominated by *ab intio* calculations of the dc polarizabilities of the clock states. Since comparison with experimentally measured ac polarizability ratios show agreement at the level of 10–20% [Fig. 5.1], we assign a conservative fractional uncertainty of 20% for the BBR shift. Equation (5.46) approximates the BBR shift in terms of the dc polarizability and a power series in $k_B T^{\text{eff}}/(\hbar\omega_{fi})$. The approximation is valid for our case since $\omega_{fi}/(2\pi c) > 8000\,\text{cm}^{-1}$, corresponding to characteristic temperatures of >11500 K much greater than T^{eff}.

The level shift of $X^1\Sigma_g^+(v, J = 0)$ due to BBR as a function of the vibrational quantum number v is shown in Fig. 5.18a, whose trend reflects that of the predicted

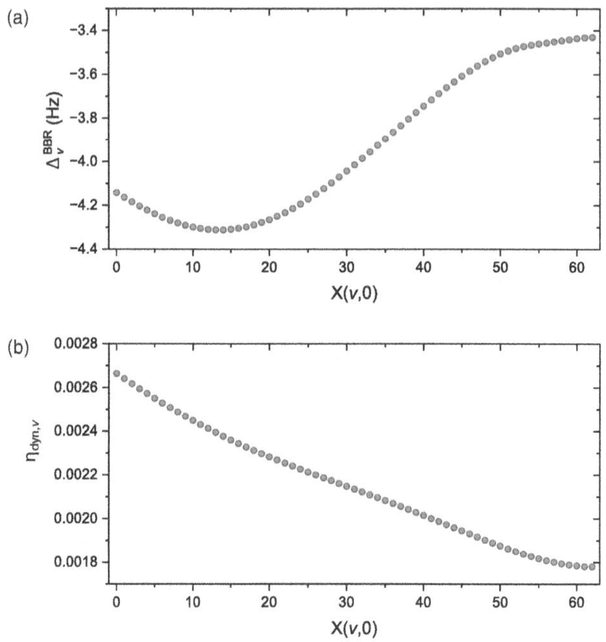

Fig. 5.18 (**a**) Calculated blackbody radiation shift of states $X^1\Sigma_g^+(v, J = 0)$ at an effective environment temperature $T^{\text{eff}} = 303$ K. For the clock states in this work, the BBR results in an observed clock frequency shifted by $\Delta f_{\text{clock}} \approx 0.7$ Hz in a room temperature environment. (b) The dynamic term $\eta_{\text{dyn,v}}$ at T^{eff}, included in Δ_v^{BBR}, contributes less than 0.5% to the shifts

dc polarizabilities of the states. The non-monotonic behavior is due to contributions from the $(4)^1\Pi_u$ potential [13]. If this prediction is correct, it poses the attractive possibility of constructing a clock (e.g., operating on $v = 0 \to v = 26$) with an exceptionally small BBR clock shift. The dynamic term contributes $< 0.5\%$ to the total BBR shift [Fig. 5.18b].

5.6.4 *Density Shift*

Due to their bosonic character, our $^{88}Sr_2$ molecules are unprotected against s-wave collisions. The one-dimensional lattice forms a series of microtraps, each with a trap volume $V = \left[2\pi k_B T/(\tilde{\omega}^2 M)\right]^{3/2}$; see Sect. 4.4.1. We investigate density-dependent shifts arising from dimer-dimer collisions by modulating the average number of molecules per lattice site ($N_{\text{mol/site}}$) at the beginning of the clock pulse. This is achieved by inserting a wait time immediately after photoassociation (PA) so that two-body collisions naturally reduce the molecule number. Fluctuations in $N_{\text{mol/site}}$ are typically $< 20\%$, and we assume equal occupancy across filled sites.

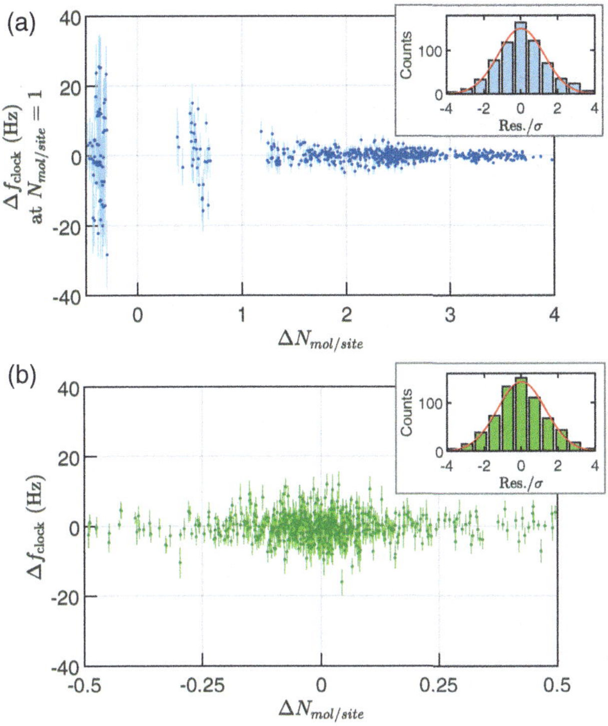

Fig. 5.19 Density shift evaluation. (**a**) Clock shifts due to molecular collisions extrapolated to operating conditions (1 molecule per lattice site, averaged over filled sites), plotted versus the change in molecule number per site used for the interleaved measurement. A single constant suffices to fit the data (0.20(10) Hz, $\chi^2_{\mathrm{red}} = 1.7$). (**b**) In the same dataset, the shift between successive resonances taken under identical experimental settings serves as a control experiment to check for technical offsets. As expected, this averages to zero (0.03(20) Hz, $\chi^2_{\mathrm{red}} = 2.0$). All statistical errors are scaled up by $\sqrt{\chi^2_{\mathrm{red}}}$. Error bars represent 1σ uncertainties. Both insets show the histogram of normalized residuals, and the solid red lines are Gaussian fits

Since both T and $\tilde{\omega}^2$ scale similarly with U_0, and the lattice intensity is stabilized, $N_{\mathrm{mol/site}}$ is a robust observable proportional to the molecular density.

Assuming linear density shifts, we scale our differential measurements to find $\Delta f_{\mathrm{clock}}$ at the normal operating value of $N_{\mathrm{mol/site}} = 1$. Figure 5.19a summarizes the measurements performed at various number differences ($\Delta N_{\mathrm{mol/site}}$) suggesting a correction of $-0.20(10)$ Hz, or $-0.63(31) \times 10^{-14}$ in fractional units, due to collisional shifts. Control measurements using spectra taken under common experimental settings do not show evidence of spurious offsets in our data [Fig. 5.19b]. It is instructive to compare the size of our density shift with similarly performing atomic clocks. The shift coefficient has a magnitude of $2.9(1.5) \times 10^{-25}$ cm^3 after normalizing by the transition frequency. This is rather similar to the analogous optical atomic clock with bosonic ^{88}Sr ($\sim 2 \times 10^{-25}$ cm^3 [53]), while being orders of magnitude smaller than in Cs ($\sim 1 \times 10^{-21}$ cm^3 [54, 55]) or Rb microwave clocks ($\sim 5 \times 10^{-23}$ cm^3 [56]).

Density effects in our clock may be suppressed due to the reactive chemistry; trimer formation releases sufficient energy to eject the products out of the trap that the spectroscopy does not probe colliding molecules. Future work may circumvent collisional shifts altogether by preparing samples with single molecule occupancy in a three-dimensional lattice or an optical tweezer array.

5.6.5 Other Clock Systematics

Effects of Magnetic Field The use of singlet and irrotational $X^1\Sigma_g^+(J = 0)$ clock states confer a high degree of insensitivity to external magnetic fields. This is an advantage of the molecular clock over the analogous optical atomic clock. Hyperfine sublevels are absent in our dimer assembled from ^{88}Sr, which has a total nuclear spin of zero. While the excited molecular states of $^{88}Sr_2$ near the intercombination line have been thoroughly studied and modeled in previous work [26, 57], an equivalent quantum chemistry calculation of higher-order Zeeman shifts of $X^1\Sigma_g^+(J = 0)$ ground states is beyond the scope of the present evaluation. Even without detailed theoretical modeling, we may experimentally investigate the extent to which our measurements are affected by magnetic field effects, including hypothetical Zeeman shifts of the clock states. We vary the applied magnetic field during clock interrogation with a lever arm of 3.2 G. The magnitude of the magnetic field was calibrated via spectroscopy of the atomic $^1S_0 \rightarrow {}^3P_1$ intercombination line, whose Zeeman coefficient is well-known [58]. The larger applied magnetic field slightly changes the photoassociation efficiency, which we partially compensate for by simultaneously varying the initial molecule number. Interleaved measurements obtain a differential shift of 0.05(41) Hz. For hypothetical shifts that scale quadratically with magnetic field strength, the measurement suggests that these contribute $< 5 \times 10^{-16}$ to the systematic uncertainty.

Magnetic fields can indirectly perturb the scanned clock lines through the excited states, such as the Raman intermediate $(1)0_u^+(J' = 1)$ state, and the $(1)1_u(J' = 1)$ state off resonantly addressed by the magic lattice. For instance, a finite magnetic field induces a quadratic Zeeman shift on the $(1)0_u^+(v' = 11, J' = 1, M_{J'} = 0)$ sublevel addressed by the polarizations of the clock lasers [26, 57]. Changes in the strength of the magnetic field, in turn, affect the common Raman detuning and the total probe light shift. Our present magnetic field stabilization is rudimentary. For the sake of argument, suppose that the bias magnetic field varied hourly by 0.1 G (a very comfortable upper bound), linear extrapolation from the above null measurement suggests that this affects the stability of the probe light shifts at $< 4 \times 10^{-16}$.

dc Stark Shift The spherically shaped stainless-steel vacuum chamber is electrically grounded, functioning as a Faraday cage. The molecules are held at a distance of approximately $R_{\text{chamb}} = 120\,\text{mm}$ from each of the fused silica viewports. We have operated the vacuum system for over a decade. Thus, we expect any stray charges on the viewports to have migrated and decayed to a negligible amount. Even if a hypothetical, improbably large voltage difference of $V_{\text{viewport}} = 20\,\text{V}$ were present between two opposite-facing viewports (giving rise to a dc electric field $\mathcal{E}_{\text{dc}} = V_{\text{viewport}}/(2R_{\text{chamb}})$), using the dc polarizability difference $\Delta\alpha_{\text{dc}} \approx 78.8$ a.u. computed with our model, the dc Stark shift[27]

$$\Delta f_{\text{clock}} = \frac{1}{2}\Delta\alpha_{\text{dc}}\mathcal{E}_{\text{dc}}^2/h, \tag{5.51}$$

multiplied by the number of such viewport pairs would amount to $< 30\,\text{mHz}$, or $< 10^{-15}$ in fractional units.

Doppler Shifts and Phase Chirps First-order Doppler shifts result from the relative motion of the lattice anti-nodes and the phases of the probe lasers. For example, this may be due to the mechanical motion of the lattice retroreflector, or phase chirps arising from the pulsing of an acousto-optic modulator (AOM) that diffracts a probe beam. Our upleg clock laser is pulsed by AOM1 before delivery to the molecules. A common solution in lattice clocks is to perform fiber noise cancellation of the probe(s) using the lattice retroreflector either as the phase reference surface or as a rigid support for a separate surface [59], which we will implement in future work. If uncompensated, AOM phase chirps can result in shifts as large as $\sim$100 mHz. We do not study phase chirps in this work. Consequently, we quote a conservative upper bound of 10^{-14} for shifts originating from this effect. The second-order Doppler shift is $< 10^{-19}$ for the typical thermal speed of our molecule.

Lattice Tunneling For the operational molecular trapping frequencies and temperature, we estimate that over 99% of molecules occupy motional quantum numbers $n < 8$, with 41% in the ground motional band ($n = 0$). A 1D lattice band structure calculation (see Sect. 2.4.3) at the operational trap depth of $U_{\text{opt}} = 487\ E_{\text{r}}$ indicates that the bandwidth of $n = 7$ is $2 \times 10^{-5}\ E_{\text{r}}$, which translates to 0.02 Hz for our molecular mass (M) and lattice wavelength (λ_{latt}). We verified that our calculation quantitatively reproduces the results of identical band structure calculations in available lattice clock literature [60, 61]. We thus quote an upper bound on Doppler-like shifts due to the delocalization of the molecular wavefunction to be $< 1\times10^{-15}$.

---

[27] Note that the expression for the dc Stark shift differs from that of the ac Stark shift (see e.g., Eq. (5.2)) by a factor of 2 because the dc shift is *not* time averaged.

This is a conservative estimate as, realistically, our horizontal lattice likely has a small but finite tilt along the direction of gravity, which should slightly suppress tunneling effects. Future upgrades to fully orient the axial direction of the lattice along gravity will further lift the Wannier state degeneracy.

Line Pulling Under operational conditions, the radial trapping frequency is 311(2) Hz, which is $10\times$ larger than the FWHM of the clock resonances. The clock and the lattice laser beams are coaligned over several meters, and the radial sidebands are not visible during normal clock operation. We model the radial sidebands as two Lorentzian peaks centered at their expected detuning from the carrier, with amplitudes equal to the size of the typical shot-to-shot signal variation. To put an upper bound on the line pulling effect, we compare the difference in the carrier line center returned by fitting a typical clock spectrum with the sum of three Lorentzians (i.e., two radial sidebands and one carrier), versus the case using just a single Lorentzian (as in Fig. 5.9c). We estimate the line pulling error to be $< 1 \times 10^{-15}$.

Scan-and-Fit Error To estimate the effect of short-term cavity flicker noise on our peak fitting, we fit a linear function to a typical time series of molecular clock lines totaling $\sim$ 3000 s (the typical duration for a single evaluation of a given systematic under interleaved clock operation). The magnitude of the linear coefficient is $<$ 10 mHz/s. For the present experiment, it takes $\sim$20 s to scan out all 15 points that make up the clock spectrum. Therefore, we estimate an upper bound for the scan-and-fit errors to be $< 6 \times 10^{-15}$.

As mentioned, the months-long average linear drift of the molecular clock line due to imperfect feedforward compensation is approximately 0.2 mHz/s. The feedforward parameters were set beforehand and unchanged during the campaign. This long-term drift would contribute a systematic offset of magnitude 1×10^{-16}, which is negligible for the current evaluation.

5.7 Absolute Frequency Evaluation

We reference all RF frequency counters and direct digital frequency synthesizers (DDS) in the experiment to a free-running rubidium microwave standard (our local timebase), as illustrated in Fig. 5.9b. Calibration of this rubidium clock is accomplished by comparing its 1 pulse-per-second (PPS) output with that of a dual-band global navigation satellite system (GNSS) receiver[28] on a time interval counter[29] (TIC).

The microprocessor within the GNSS receiver has a given clock speed (rate), thus a computational, communication, and functional dead time equal to the inverse of the clock rate. This presents an issue if the receiver has to output a pulse during this window of time. Typically, to overcome this, GNSS receivers report *quantization errors* that accompany the 1 PPS output pulses. This is the timing error of the physical output pulse compared to the ideal pulse that is synchronized to GPS time. The quantization errors vary from pulse-to-pulse and are approximately within ± 5 ns for our receiver. Each TIC measurement is corrected for its quantization error in post-process, which dramatically improves the time deviation of the measurements and the accuracy of the extracted fractional frequency offset of the rubidium timebase.

The rubidium clock serves as a flywheel oscillator to access Global Positioning System (GPS) time. In turn, GPS time is closely steered ($\lesssim 1\,\mu$s) toward UTC(USNO), one of two local realizations of Coordinated Universal Time (UTC) in the United States; the other being UTC(NIST). The timing differences between local realizations of UTC and International Atomic Time (TAI), and the calculated difference between TAI and the SI unit of the second are published in an international monthly bulletin (or report card) called the "Circular T". By definition, the scale intervals of UTC and TAI are equivalent.

The full frequency chain can be algebraically written as

$$\frac{f_{\mathrm{clock}}}{f_{\mathrm{SI}}} = \frac{f_{\mathrm{clock}}}{f_{\mathrm{Rb}}} \times \frac{f_{\mathrm{Rb}}}{f_{\mathrm{GPS}}} \times \frac{f_{\mathrm{GPS}}}{f_{\mathrm{UTC(USNO)}}} \times \frac{f_{\mathrm{UTC(USNO)}}}{f_{\mathrm{TAI}}} \times \frac{f_{\mathrm{TAI}}}{f_{\mathrm{SI}}}, \tag{5.52}$$

where $f_{\mathrm{SI}} \equiv 1$ Hz is the SI unit of frequency such that $f_{\mathrm{clock}}/f_{\mathrm{SI}}$ is the numerical value of the absolute frequency of the molecular clock. $f_{\mathrm{clock}}/f_{\mathrm{Rb}}$ is the molecular clock frequency relative to the rubidium clock that serves as our lab timebase (calcu-

[28] The dual band antenna (Tallysman 3972XF) that receives the broadcasted GNSS messages from the satellites is lightning protected and mounted on the roof of Pupin Hall with a clear view of the sky. The signals are passed through an in-line amplifier and transmitted 6 or 7 floors down over $\sim$ 200 ft of compatible coaxial cable from the roof to the lab, where the GNSS receiver (ZED-F9T) is housed. A GNSS receiver solves the navigation equations to obtain the geographical position of the antenna and the timing difference between the receiver's clock and the GNSS constellation's time scale (here chosen to be GPS). To operate the receiver in "Timing" mode, a survey of the antenna's position is made over 24 hours using the accompanied manufacturer software, after which the spatial coordinates are automatically saved. The survey suggests that the antenna is 87.2 m above Earth's geoid (essentially mean sea level). If the position of the antenna is changed, this calibration survey will have to be repeated. Note that the receiver is agnostic of the distance (and indeed the height difference) between it and the antenna. Nevertheless, international timescales are engineered to closely approximate the proper time kept by a stationary clock on the geoid, so we only need to account for gravitational time dilation due to the height of the molecular clock apparatus relative to the geoid.

[29] Pendulum CNT-91.

lated via Eq. (5.35) using the molecular clock line centers, the counted repetition rate of the frequency comb, and the molecular clock systematic corrections). $f_{\mathrm{Rb}}/f_{\mathrm{GPS}}$ is the frequency of the rubidium clock relative to GPS time. GPS time is closely steered toward UTC(USNO), and $f_{\mathrm{GPS}}/f_{\mathrm{UTC(USNO)}}$ is the frequency of GPS time relative to UTC(USNO). Finally, $f_{\mathrm{UTC(USNO)}}/f_{\mathrm{TAI}}$ is the frequency of UTC(USNO) relative to TAI, and $f_{\mathrm{TAI}}/f_{\mathrm{SI}}$ is the frequency of TAI relative to the SI second.

The molecular clock is operated intermittently due to its complexity and the practical challenges of our present experimental apparatus. As such, the clock was not continuously phase-linked with the SI second. To address this, we expand Eq. (5.52) as

$$\frac{f_{\mathrm{clock}}}{f_{\mathrm{SI}}} = \frac{f_{\mathrm{clock}}}{f_{\mathrm{Rb}}(\mathcal{T}_1)} \times \frac{f_{\mathrm{Rb}}(\mathcal{T}_1)}{f_{\mathrm{GPS}}(\mathcal{T}_1)} \times \frac{f_{\mathrm{GPS}}(\mathcal{T}_1)}{f_{\mathrm{GPS}}(\mathcal{T}_2)} \times \frac{f_{\mathrm{GPS}}(\mathcal{T}_2)}{f_{\mathrm{UTC(USNO)}}(\mathcal{T}_2)} \times \frac{f_{\mathrm{UTC(USNO)}}(\mathcal{T}_2)}{f_{\mathrm{TAI}}(\mathcal{T}_2)} \times \frac{f_{\mathrm{TAI}}(\mathcal{T}_2)}{f_{\mathrm{SI}}}. \tag{5.53}$$

Here, $\mathcal{T}_1$ is the typical up time of the molecular clock corresponding to one measurement trial [Fig. 5.22], $\mathcal{T}_2 = 1$ month corresponds to the time period for the publication of the Circular T [62], and the bracketed time explicitly states the duration over which the given frequency is averaged. We have assumed that the SI second and molecular clock frequency are unchanging in time.

Over the length of the campaign, the scale intervals of GPS time, UTC(USNO), and International Atomic Time (TAI) differed by $\lesssim 10^{-15}$, and the daily fractional changes in the frequency of GPS time relative to TAI are $\lesssim 10^{-14}$ [63]. Therefore, for the present study, except for $f_{\mathrm{Rb}}(\mathcal{T}_1)/f_{\mathrm{GPS}}(\mathcal{T}_1)$ and $f_{\mathrm{clock}}/f_{\mathrm{Rb}}(\mathcal{T}_1)$, all other ratios contribute a negligible uncertainty and may be assumed to be unity. This includes the extrapolation ratio, $f_{\mathrm{GPS}}(\mathcal{T}_1)/f_{\mathrm{GPS}}(\mathcal{T}_2)$, which is the frequency of GPS time during the molecular clock up times versus the frequency of GPS time broadcasted by the constellation over a month.

Each TIC measurement is started by the rising edge of the 1 PPS from the rubidium clock and stopped by the rising edge of the 1 PPS from the GNSS receiver. Thus, the instantaneous fractional frequency *offset* of the rubidium clock relative to the frequency of GPS time, $r = [f_{\mathrm{Rb}}/f_{\mathrm{GPS}}] - 1$, is quantified by the instantaneous slope of the logged TIC measurements as a function of elapsed time. This measurement is susceptible to noise in the satellite link, as well as the instabilities of GPS time and the rubidium clock. Comparisons with an identical, independent free-running rubidium clock indicate that the rubidium clock reaches an instability flicker floor of approximately $3 \times 10^{-13} \equiv \sigma_{\mathrm{Rb}}$ after $\sim 5 \times 10^3$ s of averaging time (comparable to typical durations of $\mathcal{T}_1$), but worsens to $\sim 10^{-12}$ for time periods over 24 hours. This poses a conundrum, because at least 24 hours of continuous averaging is typically required to achieve an inaccuracy and instability of $< 10^{-13}$ using one-way GPS time transfer [64], but the rubidium clock is not a good flywheel on these time scales.

Rubidium microwave standards are more susceptible to unpredictable environmental perturbations than, for instance, hydrogen masers [65], making it challenging

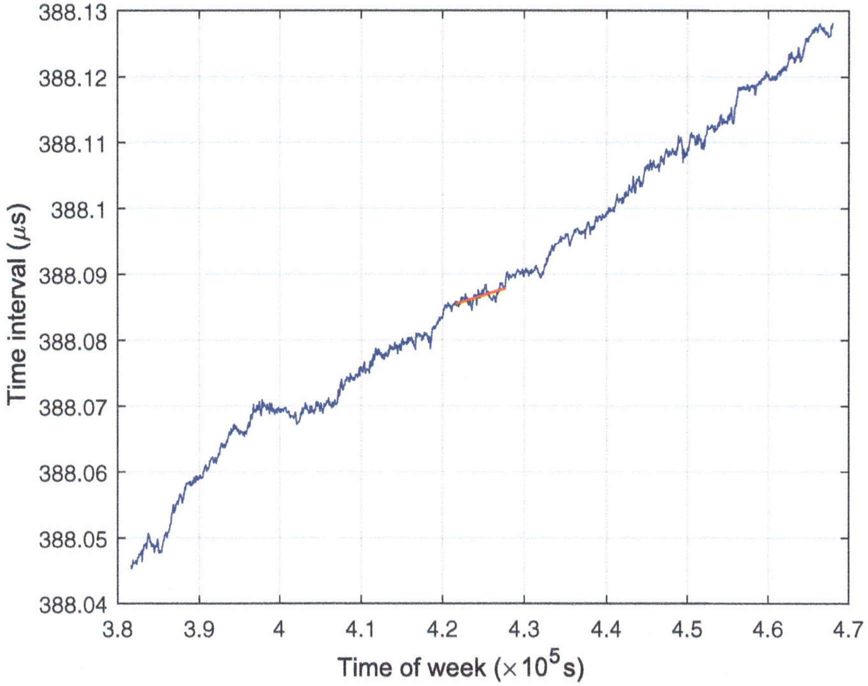

Fig. 5.20 Example post-processed time interval counter (TIC) log, taken on GPS week 2215, for the molecular clock run between MJD 59753.5–59754.5. The TIC measurements are started by the rising edge of the 1 PPS from the rubidium clock and stopped by the rising edge of the 1 PPS from the GNSS receiver. The quantization errors reported by the GNSS receiver are simultaneously logged and corrected point-by-point in post-process, resulting in the blue points. The red solid line is a linear fit to the segment that coincides with the molecular clock up time. A positive (negative) slope implies that the rubidium clock runs faster (slower) than GPS; i.e., "10 MHz" frequency according to the local timebase is actually slightly larger (smaller) if it were measured against the international system of units. Therefore, all frequencies referenced to the rubidium timebase would have been erroneously measured to be smaller (larger), and a positive (negative) correction is necessary. Triggering on the rising edge of the GNSS receiver (configured to have rising pulse polarity) is preferred because its pulse width is large ($\sim$100 ms), and consequently, its falling edge may be less stable. Note that the ZED-F9T receiver reports quantization errors for the *succeeding* pulses. In addition, there are inherent latencies ($\lesssim$ 100 ms) in the GNSS receiver timing computations, as well as the communication between the receiver and the logging computer. Thus, to avoid misidentifying the reported quantization error with the corresponding TIC measurement (which the logging computer queries upon receipt of a valid report from the GNSS receiver), we make the time intervals on the TIC short ($\sim$ 100 μs) prior to logging by adjusting the placement of the rubidium clock pulses. To improve the visual fidelity of the figure in print, the blue points are plotted here sparsely (every 60 points). Note that all analyses and conclusions are drawn from the complete dataset

to construct a reliable noise model. Therefore, for every trial, we operationally extract $r = [f_{\text{Rb}}(\mathcal{T}_1)/f_{\text{GPS}}(\mathcal{T}_1)] - 1$ through linear fitting of the TIC measurements as a function of elapsed time, restricting the fits to the durations coinciding with the up time segments of the molecular clock. This process is illustrated in Fig. 5.20.

We judged a detailed characterization of the satellite link to be beyond the scope of the present work. Geometric multipath effects and the diurnal variation in the ionosphere may introduce a systematic offset ($\approx 2 \times 10^{-13} \equiv \sigma_{\mathrm{GPS,sys}}$), since a majority of the molecular clock up times were in the afternoon.

We estimate the fractional uncertainty of the extracted values of r to be $\sqrt{\sigma^2_{\mathrm{Rb}} + \sigma^2_{\mathrm{GPS,tot}}}$, where $\sigma^2_{\mathrm{GPS,tot}} = \sigma^2_{\mathrm{GPS,stat}} + \sigma^2_{\mathrm{GPS,sys}}$ and $\sigma_{\mathrm{GPS,stat}} \approx 10^{-13} \times \sqrt{86400/\mathcal{T}_1\,[s]}$. The uncertainty from linear fitting is an order of magnitude smaller. Occasionally, we manually realigned the rubidium clock frequency relative to GPS if it exceeded a fractional offset of 1×10^{-11}. This is not done during the molecular clock up times, nor within 24 hours of those segments to let the rubidium clock settle. For every trial, we add a unique frequency correction $r \times [f_{\mathrm{GPS}}(\mathcal{T}_1)/f_{\mathrm{GPS}}(\mathcal{T}_2)] \times [f_{\mathrm{clock}}/f_{\mathrm{Rb}}(\mathcal{T}_1)]$ to $f_{\mathrm{clock}}/f_{\mathrm{Rb}}(\mathcal{T}_1)$, obtaining the absolute frequency of the molecular clock.

Figure 5.21 shows the results of the measurement campaign, consisting of 10 trials performed on separate days. For clarity, we reiterate that all known systematic corrections have been applied, including the frequency corrections due to the daily systematic offset of the rubidium clock. Each absolute clock frequency measurement trial utilizes clock lines taken under operational conditions [Fig. 5.22]. For experimental ease, the clock lasers and frequency comb are not actively steered toward the molecular resonance in real time. Instead, we repeatedly scan the clock transition to obtain a time series of the line centers while counting the frequency comb's repetition rate, f_{RR}, against the rubidium timebase on a zero dead time frequency counter (see Sect. 3.1.2). Figure 3.6b suggests that the combined drift of the reference cavity and the rubidium timebase is $\sim 10^{-13}$ for averaging times $< 10^3$ s. This means that we can meaningfully average, in post-process, the measurements of f_{RR} over a window of time $\lesssim 10^3$ s to achieve better statistics beyond a single counter measurement. To preclude the residual cavity drift from biasing the averaged repetition rates, we quite conservatively use a 5-minute window; i.e., we average $5 \times 60 + 1 = 301$ consecutive counter measurements of f_{RR} for each point in Fig. 5.22. The window is centered on the timestamp closest to the scan over the corresponding molecular clock line. The molecular clock systematics are controlled at the level quoted in Table 5.3. The probe light shifts were evaluated every trial to account for potential daily variations in probe laser beam pointing. We determine the elevation of our apparatus to be $H = 51(5)$ m above mean sea level. Clocks deeper in a gravitational potential run slower [66–69]; i.e., $\Delta f_{\mathrm{clock}}/f_{\mathrm{clock}} = gH/c^2$, where g is the local gravitational acceleration. Thus, the redshift correction to f_{clock} is $-0.18(2)$ Hz.

A weighted average over the 10 trials yields the absolute frequency of the $^{88}\mathrm{Sr}_2$ vibrational clock to be

$$\boxed{f_{\mathrm{clock}} = 31\,825\,183\,207\,592.8(5.1)\,\mathrm{Hz},} \tag{5.54}$$

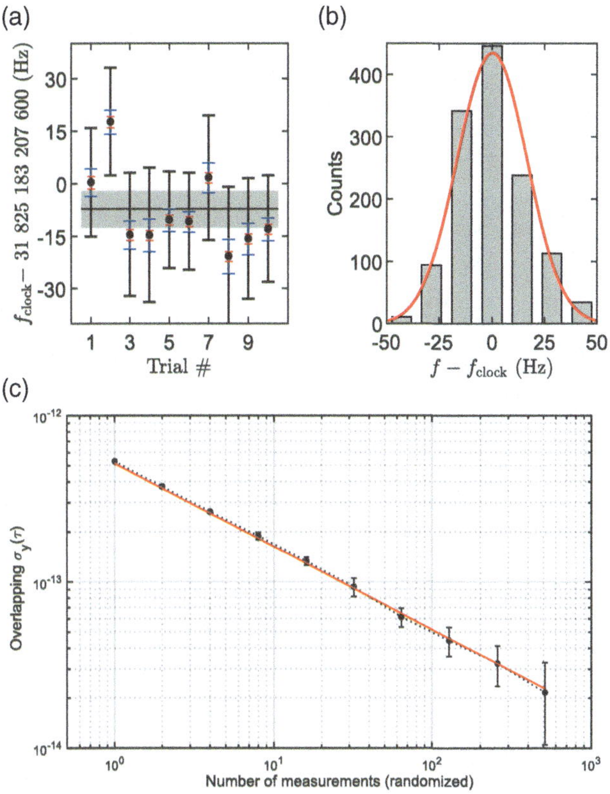

Fig. 5.21 (**a**) Absolute frequency of the clock transition measured over 10 trials (filled black circles) with all known frequency offsets corrected, including that of the local rubidium timebase (see main text for details). Blue error bars are 1σ statistical uncertainties, dominated by the determination of the comb repetition rate rather than the stability of the scanned molecular clock lines. Red error bars are 1σ systematic uncertainties due to the molecular clock only (see Table 5.3). Black error bars are 1σ total uncertainties, where the uncertainties of the local timebase calibrations are added in quadrature with the statistical and molecular clock systematic uncertainties. The black horizontal line shows the weighted average ($\chi^2_{\rm red} = 0.5$), and the shaded gray area shows the associated $\pm 1\sigma$ standard error of the mean. (**b**) Histogram of all clock frequency measurements in the 10 trials, relative to the weighted average of $f_{\rm clock}$. The solid red line is a Gaussian fit to the histogram, which serves as a guide to the eye. (**c**) Overlapping Allan deviation of all clock frequency measurements (randomized). See Fig. 5.22 for the expanded dataset

where the quoted uncertainty is the revised standard error of the mean. In future work, we expect to reduce the uncertainty of our local timebase calibration to the same level as the molecular clock systematics (or better) by upgrading to a standard with intrinsically lower instability (e.g., cesium beam standard, or hydrogen maser) and utilizing two-way time transfer schemes (e.g., GPS carrier phase) with a national timing institute.

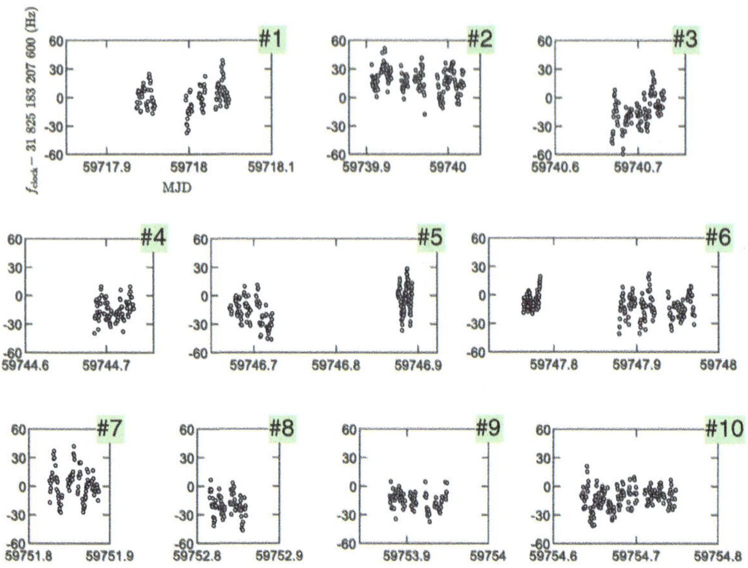

Fig. 5.22 Expanded dataset for the absolute clock frequency measurement campaign. For experimental ease, the clock lasers and frequency comb are not actively steered toward the molecular resonance in real time. Instead, each measurement (filled gray circle) is calculated from a single scan of the molecular clock transition under operational conditions, and the repetition rate of the frequency comb is counted with a zero dead time frequency counter (1 s gate time). The shot-to-shot noise in the counted repetition rates is lowered by taking their average over a short window of time centered on the timestamp of the corresponding molecular clock measurement in post-process. Since the averaging windows are conservatively chosen to be short (5 min), the statistical error on each point is typically ~40 Hz. The data is plotted against the time of measurement (Modified Julian Date, MJD). The frequency values have been corrected for all systematic errors in Table 5.3, the gravitational redshift, and the rubidium timebase error. Error bars are not shown for visual clarity. Trial numbers on the top right corner of each plot correspond to those in Fig. 5.21a. The histogram and Allan deviation of all measurements is shown in Figs. 5.21b, c. The axes labels are displayed in the plot for Trial 1 (top left), and are the same for the remaining plots

To our knowledge, f_{clock} represents one of the most accurately measured pure molecular vibrational frequencies to date. The fractional uncertainty (1.6×10^{-13}) is on par with that of the unidentified rovibrational interval in OsO_4 near the R(10) (00^01)–(10^00) emission line of the $^{12}C^{16}O_2$ laser. This absorption line in OsO_4 is a secondary representation of the SI second [70], and was compared directly against a primary cesium standard by stabilizing a CO_2 laser to the specific saturated absorption feature of OsO_4 in a high-finesse cavity [71, 72].

Accurate terahertz (THz) frequency metrology is vital for many applications, yet few frequency standards currently exist in the THz band [70, 73]. Our prototype clock is the first THz frequency standard based on ultracold molecules and can generate stable radiation at 9.4 μm via photomixing [74, 75]. Alternatively, transitions in heteronuclear isotopologues (e.g., $^{86}Sr^{88}Sr$, $^{84}Sr^{88}Sr$, $^{86}Sr^{84}Sr$) could be driven directly with quantum cascade lasers [76, 77].

5.8 The Dissociation Energy of $^{88}Sr_2$

f_{clock} is the unperturbed *vibrational splitting* of $X(62, 0)$ and $X(0, 0)$. Therefore, the sum of f_{clock} with the binding energy of the least bound state $X(62, 0)$, $|E^{X0_g^+}_{v=62,J=0}|$, yields the dissociation energy ($D_0 \equiv |E^{X0_g^+}_{v=0,J=0}|$) of our molecule with respect to the $^1S_0 + {}^1S_0$ threshold,

$$D_0 = f_{\text{clock}} + |E^{X0_g^+}_{v=62,J=0}|. \tag{5.55}$$

While the analogous least bound vibrational states of $^{84}Sr_2$ and $^{86}Sr_2$ are known with sub-kHz uncertainties [78, 79], the current best measurement for $^{88}Sr_2$ is at the kHz-level [58]. Nevertheless, taking the binding energy of $X(62, 0)$ to be 136.6447(50) MHz from Ref. [58], which was determined using two-photon dissociation, we find

$$\boxed{D_0(^{88}Sr_2) = 31\,825\,319\,852(5)\text{ kHz} = 1\,061.578\,402\,09(17)\text{ cm}^{-1} \times c.} \tag{5.56}$$

This is an improvement by 5 orders of magnitude over the previously reported value for Sr_2 in available literature [16], and sets a new accuracy record for the determination of a molecular dissociation energy (1.6×10^{-10} fractional uncertainty). To list a few competitive results, dissociation energies have been reported with fractional uncertainties of 4.4×10^{-10} for $^{87}Rb^{133}Cs$ [80], 6.9×10^{-10} for ortho-H_2 [81], 8.6×10^{-10} for para-H_2 [82], and 7.1×10^{-10} for ortho-D_2 [83].

5.9 Conclusion and Outlook

Exerting precise quantum control over the rovibronic states of molecules was considered a grand challenge nearly two decades ago. Nevertheless, quantum science reached a level of maturity in the intervening years that has enabled our experiments with diatomic strontium, among others.

The work in this thesis arose from the amalgamation of various disciplines, namely ultracold gases, quantum chemistry, laser physics, atomic clock-making, and frequency metrology. Just as we anticipate and recognize individual musical notes that flow harmoniously from an arpeggiated chord, we can triangulate seemingly disparate ideas and sense their connections as part of a broader picture, even when the future is yet to come.

Molecular spectroscopy is increasingly appreciated as a fertile ground in the search for new physics. Besides contributing to the global network of clocks and opening opportunities for bridging the THz metrology gap, molecular clocks are also poised to broaden the scope of fundamental measurements [84, 85]. These include tests of molecular quantum electrodynamics, fifth forces at nanometer scales, the time variation of the fundamental constants, and various models of ultralight dark matter.

Both accuracy and precision are essential metrics for a metrological platform, and they often advance in lockstep. Going beyond the work in this thesis, realizing a vibrational lattice clock with systematic uncertainty and instability at (or below) the 10^{-15} level will require several technical advancements that overcome the limitations in our present apparatus. If history is any guide, these advances may extend the capabilities of quantum computing and quantum simulation experiments that share the challenge of engineering rovibronic molecular states with long quantum coherence times that are robust against external environmental influences.

It is still very early days for the vibrational molecular clock. It took over a decade and the efforts of multiple laboratories worldwide for atomic lattice clocks to advance to their record performance (currently with systematic uncertainties at the low 10^{-18} [47, 86], and instabilities at the 10^{-21} level [68]). From this perspective, molecular clocks are on an upward trajectory.

References

1. Bonin, K.D., Kresin, V.V.: Electric-Dipole Polarizabilities of Atoms, Molecules, and Clusters. World Scientific, Singapore (1997)
2. Le Kien, F., Schneeweiss, P., Rauschenbeutel, A.: Dynamical polarizability of atoms in arbitrary light fields: general theory and application to cesium. Eur. Phys. J. D **67**, 1–16 (2013)
3. Davydkin, V., Ovsiannikov, V.: The hyperpolarisability of an excited atom. J. Phys. B: At. Mol. Phys. **19**, 2071 (1986)
4. Manakov, N.L., Ovsiannikov, V.D., Rapoport, L.P.: Atoms in a laser field. Phys. Rep. **141**, 320–433 (1986)
5. Katori, H., Takamoto, M., Pal'Chikov, V., Ovsiannikov, V.: Ultrastable optical clock with neutral atoms in an engineered light shift trap. Phys. Rev. Lett. **91**, 173005 (2003)
6. Zare, R.N., Harter, W.G.: Angular Momentum: Understanding Spatial Aspects in Chemistry and Physics. Wiley-Interscience, Hoboken (1988)
7. Mitroy, J., Safronova, M.S., Clark, C.W.: Theory and applications of atomic and ionic polarizabilities. J. Phys. B: At. Mol. Opt. Phys. **43**, 202001 (2010)
8. Safronova, M.S., Porsev, S.G., Safronova, U.I., Kozlov, M.G., Clark, C.W.: Blackbody-radiation shift in the Sr optical atomic clock. Phys. Rev. A **87**, 012509 (2013)
9. Caldwell, L., Tarbutt, M.: Sideband cooling of molecules in optical traps. Phys. Rev. Res. **2**, 013251 (2020)
10. McGuyer, B.H., et al.: Precise study of asymptotic physics with subradiant ultracold molecules. Nat. Phys. **11**, 32–36 (2015)
11. Barber, Z.W., et al.: Optical lattice induced light shifts in an Yb atomic clock. Phys. Rev. Lett. **100**, 103002 (2008)
12. Kobayashi, T., et al.: Uncertainty evaluation of an ^{171}Yb optical lattice clock at NMIJ. IEEE Trans. Ultrason. Ferroelectr. Freq. Control **65**, 2449–2458 (2018)
13. Majewska, I.: Theoretical description of ultracold strontium molecules in an optical lattice: control of photodissociation and interpretation of molecular clock experiments. PhD thesis (University of Warsaw, Warsaw, 2021)
14. Skomorowski, W., Pawłowski, F., Koch, C.P., Moszynski, R.: Rovibrational dynamics of the strontium molecule in the $A^1\Sigma_u^+$, $c^3\Pi_u$, and $a^3\Sigma_u^+$ manifold from state-of-the-art ab initio calculations. J. Chem. Phys. **136**, 194306 (2012)

15. Werner, H., et al.: MOLPRO, version 2019.2, a package of ab initio programs (2019)
16. Stein, A., Knöckel, H., Tiemann, E.: The ^{1}S + ^{1}S asymptote of Sr_2 studied by Fourier-transform spectroscopy. Eur. Phys. J. D **57**, 171–177 (2010)
17. Stein, A., Knöckel, H., Tiemann, E.: The states $1^1\Sigma_u^+$, $1^1\Pi_u$and $2^1\Sigma_u^+$ of Sr_2 studied by Fourier-transform spectroscopy. Eur. Phys. J. D **64**, 227–238 (2011)
18. Leung, K.H., et al.: Transition strength measurements to guide magic wavelength selection in optically trapped molecules. Phys. Rev. Lett. **125**, 153001 (2020)
19. Ushijima, I., Takamoto, M., Katori, H.: Operational magic intensity for Sr optical lattice clocks. Phys. Rev. Lett. **121**, 263202 (2018)
20. Porsev, S.G., Ludlow, A.D., Boyd, M.M., Ye, J.: Determination of Sr properties for a high-accuracy optical clock. Phys. Rev. A **78**, 032508 (2008)
21. Tiesinga, E., Williams, C.J., Julienne, P.S.: Photoassociative spectroscopy of highly excited vibrational levels of alkali-metal dimers: green-function approach for eigenvalue solvers. Phys. Rev. A **57**, 4257–4267 (1998)
22. Kondov, S.S., et al.: Molecular lattice clock with long vibrational coherence. Nat. Phys. **15**, 1118–1122 (2019)
23. Sansonetti, J., Nave, G.: Wavelengths, transition probabilities, and energy levels for the spectrum of neutral strontium (Sr I). J. Phys. Chem. Ref. Data **39**, 033103 (2010)
24. Lee, C.-H.: Quantum metrology with a molecular lattice clock and state-selected photodissociation of ultracold molecules. PhD thesis (Columbia University, New York, 2020)
25. Fitch, N.J., Tarbutt, M.R.: Laser-cooled molecules. Adv. At. Mol. Opt. Phys. **70**, 157–262 (2021)
26. McGuyer, B., et al.: Control of optical transitions with magnetic fields in weakly bound molecules. Phys. Rev. Lett. **115**, 053001 (2015)
27. Gehm, M., O'hara, K., Savard, T., Thomas, J.: Dynamics of noise-induced heating in atom traps. Phys. Rev. A **58**, 3914 (1998)
28. Skomorowski, W., Moszynski, R., Koch, C.P.: Formation of deeply bound ultracold Sr_2 molecules by photoassociation near the $^1S + ^3P_1$intercombination line. Phys. Rev. A **85**, 043414 (2012)
29. Ma, L.-S., Jungner, P., Ye, J., Hall, J.L.: Delivering the same optical frequency at two places: accurate cancellation of phase noise introduced by an optical fiber or other time-varying path. Opt. Lett. **19**, 1777–1779 (1994)
30. Rauf, B., Vélez López, M., Thoumany, P., Pizzocaro, M., Calonico, D.: Phase noise cancellation in polarisation-maintaining fibre links. Rev. Sci. Instrum. **89**, 033103 (2018)
31. Alnis, J., Matveev, A., Kolachevsky, N., Udem, T., Hänsch, T.: Subhertz linewidth diode lasers by stabilization to vibrationally and thermally compensated ultralow-expansion glass Fabry-Pérot cavities. Phys. Rev. A **77**, 053809 (2008)
32. Bevington, P.R., Robinson, D.K.: Data Reduction and Error Analysis for the Physical Sciences, 3rd edn. McGraw-Hill Higher Education, New York (2003)
33. Richter, P.H.: Estimating errors in least-squares fitting. The Telecommunications and Data Acquisition Report (1995)
34. Gallant, A.R.: Nonlinear regression. Am. Stat. **29**, 73–81 (1975)
35. Serber, G.A., Wild, C.J.: Nonlinear Regression. Wiley, New York (1989)
36. Nicholson, T.L.: A new record in atomic clock performance PhD thesis (University of Colorado Boulder, Boulder, 2015)
37. Fasano, R., et al.: Characterization and suppression of background light shifts in an optical lattice clock. Phys. Rev. Appl. **15**, 044016 (2021)
38. Brown, R.C., et al.: Hyperpolarizability and operational magic wavelength in an optical lattice clock. Phys. Rev. Lett. **119**, 253001 (2017)
39. Beloy, K., et al.: Modeling motional energy spectra and lattice light shifts in optical lattice clocks. Phys. Rev. A **101**, 053416 (2020)
40. Jackson, S., Vutha, A.C.: Magic polarization for cancellation of light shifts in two-photon optical clocks. Phys. Rev. A **99**, 063422 (2019)

41. Yudin, V., et al.: Generalized autobalanced Ramsey spectroscopy of clock transitions. Phys. Rev. Appl. **9**, 054034 (2018)
42. Zanon-Willette, T., de Clercq, E., Arimondo, E.: Probe light-shift elimination in generalized hyper-Ramsey quantum clocks. Phys. Rev. A **93**, 042506 (2016)
43. Zanon-Willette, T., et al.: Cancellation of stark shifts in optical lattice clocks by use of pulsed Raman and electromagnetically induced transparency techniques. Phys. Rev. Lett. **97**, 233001 (2006)
44. Hobson, R., et al.: Modified hyper-Ramsey methods for the elimination of probe shifts in optical clocks. Phys. Rev. A **93**, 010501 (2016)
45. Porsev, S.G., Derevianko, A.: Multipolar theory of blackbody radiation shift of atomic energy levels and its implications for optical lattice clocks. Phys. Rev. A **74**, 020502 (2006)
46. Beloy, K., et al.: Atomic clock with 1×10^{-18} room-temperature blackbody stark uncertainty. Phys. Rev. Lett. **113**, 260801 (2014)
47. Bothwell, T., et al.: JILA SrI optical lattice clock with uncertainty of 2.0×10^{-18}. Metrologia **56**, 065004 (2019)
48. American Society of Heating, Refrigerating and Air-Conditioning Engineers. ASHRAE Hand- book: Fundamentals (2009)
49. Barnes, B., Forsythe, W., Adams, E.: The total emissivity of various materials at 100–500°C. J. Opt. Soc. Am. **37**, 804–807 (1947)
50. Wieting, T., DeRosa, J.: Effects of surface condition on the infrared absorptivity of 304 stainless steel. J. Appl. Phys. **50**, 1071–1078 (1979)
51. Wittenberg, A.: Total hemispherical emissivity of sapphire. J. Opt. Soc. Am. **55**, 432–435 (1965)
52. Joint Committee for Guides in Metrology. Evaluation of measurement data—Guide to the expression of uncertainty in measurement. JCGM **100**, 1–116 (2008)
53. Lisdat, C., Winfred, J.V., Middelmann, T., Riehle, F., Sterr, U.: Collisional losses, decoherence, and frequency shifts in optical lattice clocks with bosons. Phys. Rev. Lett. **103**, 090801 (2009)
54. Gibble, K., Chu, S.: Laser-cooled Cs frequency standard and a measurement of the frequency shift due to ultracold collisions. Phys. Rev. Lett. **70**, 1771 (1993)
55. Dos Santos, F.P., et al.: Controlling the cold collision shift in high precision atomic interferometry. Phys. Rev. Lett. **89**, 233004 (2002)
56. Sortais, Y., et al.: Cold collision frequency shifts in a ^{87}Rb atomic fountain. Phys. Rev. Lett. **85**, 3117 (2000)
57. McGuyer, B., et al.: Nonadiabatic effects in ultracold molecules via anomalous linear and quadratic Zeeman shifts. Phys. Rev. Lett. **111**, 243003 (2013)
58. McDonald, M.: High Precision Optical Spectroscopy and Quantum State Selected Photodissociation of Ultracold $^{88}Sr_2$ Molecules in an Optical Lattice. Springer, Berlin (2017)
59. Falke, S., Misera, M., Sterr, U., Lisdat, C.: Delivering pulsed and phase stable light to atoms of an optical clock. Appl. Phys. B **107**, 301–311 (2012)
60. Lemonde, P., Wolf, P.: Optical lattice clock with atoms confined in a shallow trap. Phys. Rev. A **72**, 033409 (2005)
61. Falke, S., et al.: A strontium lattice clock with 3×10^{-17} inaccuracy and its frequency. New J. Phys. **16**, 073023 (2014)
62. Arias, E.F., Panfilo, G., Petit, G.: Timescales at the BIPM. Metrologia **48**, S145–S153 (2011)
63. Bureau International des Poids et Mesures (BIPM). Circular T 414 (June 2022)
64. Lombardi, M.A., Nelson, L.M., Novick, A.N., Zhang, V.S.: Time and frequency measurements using the global positioning system. Cal Lab: Int. J. Metrol. **8**, 26–33 (2001)
65. Marlow, B.L.S., Scherer, D.R.: A review of commercial and emerging atomic frequency standards. IEEE Trans. Ultrason. Ferroelectr. Freq. Control **68**, 2007–2022 (2021)
66. Feynman, R.P., Leighton, R.B., Sands, M.: The Feynman Lectures on Physics, Vol. I: The New Millennium Edition: Mainly Mechanics, Radiation, and Heat. Basic Books, New York (2011)

67. Takamoto, M., et al.: Test of general relativity by a pair of transportable optical lattice clocks. Nat. Photon **14**, 411–415 (2020)
68. Bothwell, T., et al.: Resolving the gravitational redshift across a millimetre-scale atomic sample. Nature **602**, 420–424 (2022)
69. Zheng, X., et al.: Differential clock comparisons with a multiplexed optical lattice clock. Nature **602**, 425–430 (2022)
70. Riehle, F., Gill, P., Arias, F., Robertsson, L.: The CIPM list of recommended frequency standard values: guidelines and procedures. Metrologia **55**, 188 (2018)
71. Daussy, C., Ducos, F., Rovera, G., Acef, O.: Performances of OsO_4 stabilized CO_2 lasers as optical frequency standards near 29 THz. IEEE Trans. Ultrason. Ferroelectr. Freq. Control **47**, 518–521 (2000)
72. Rovera, G.D., Acef, O.: Frequency Measurement and Control, pp. 249–272. Springer, Berlin (2001)
73. Shelkovnikov, A., Butcher, R.J., Chardonnet, C., Amy-Klein, A.: Stability of the proton- to-electron mass ratio. Phys. Rev. Lett. **100**, 150801 (2008)
74. Preu, S., Döhler, G., Malzer, S., Wang, L., Gossard, A.: Tunable, continuous-wave terahertz photomixer sources and applications. J. Appl. Phys. **109**, 4 (2011)
75. Hindle, F., et al.: Widely tunable THz synthesizer. Appl. Phys. B **104**, 763–768 (2011)
76. Bartalini, S., et al.: Frequency-comb-assisted terahertz quantum cascade laser spectroscopy. Phys. Rev. X **4**, 021006 (2014)
77. Consolino, L., Cappelli, F., de Cumis, M.S., De Natale, P.: QCL-based frequency metrology from the mid-infrared to the THz range: a review. Nanophotonics **8**, 181–204 (2019)
78. Stellmer, S., Pasquiou, B., Grimm, R., Schreck, F.: Creation of ultracold Sr_2 molecules in the electronic ground state. Phys. Rev. Lett. **109**, 115302 (2012)
79. Aman, J., et al.: Photoassociative spectroscopy of a halo molecule in ^{86}Sr. Phys. Rev. A **98**, 053441 (2018)
80. Molony, P.K., et al.: Measurement of the binding energy of ultracold ^{87}Rb^{133}Cs molecules using an offset-free optical frequency comb. Phys. Rev. A **94**, 022507 (2016)
81. Cheng, C.-F., et al.: Dissociation energy of the hydrogen molecule at 10^{-9} accuracy. Phys. Rev. Lett. **121**, 013001 (2018)
82. Beyer, M., et al.: Determination of the interval between the ground states of para- and ortho H_2. Phys. Rev. Lett. **123**, 163002 (2019)
83. Hussels, J., et al.: Improved ionization and dissociation energies of the deuterium molecule. Phys. Rev. A **105**, 022820 (2022)
84. Safronova, M., et al.: Search for new physics with atoms and molecules. Rev. Mod. Phys. **90**, 025008 (2018)
85. Mitra, D., Leung, K.H., Zelevinsky, T.: Quantum control of molecules for fundamental physics. Phys. Rev. A **105**, 040101 (2022)
86. McGrew, W., et al.: Atomic clock performance enabling geodesy below the centimetre level. Nature **564**, 87–90 (2018)

The manufacturer's authorised representative in the EU is Springer Nature Customer Service Centre GmbH, Europaplatz 3, 69115 Heidelberg, Germany. If you have any concerns regarding our products, please contact ProductSafety@springernature.com

Printed and bound by CPI Group (UK) Ltd, Croydon, CR0 4YY
01/10/2026
02225875-0016